200 Science Experiments

Lifestyle Publishing

© Copyright 2015

All Rights Reserved.

Table of Contents

I. CARBON DIOXIDE EXPERIMENTS

1) BAKING SODA & VINEGAR VOLCANO

This quick and easy experiment uses baking soda and vinegar to make an explosive chemical reaction! Just be sure to have some towels ready for cleanup.

Materials you'll need:

- Baking Soda
- Vinegar (any kind will do)
- A large container
- Paper towels or a cloth (for cleanup)
- Optional: Red food coloring

What to do:

1. Place some of the baking soda into your container.
2. Pour in some of the vinegar
3. Stand back and watch!

What's happening?

Baking soda (aka sodium bicarbonate) is a base, while vinegar (aka acetic acid) is an acid. When combined together, they react to form *carbonic acid* which is very unstable. It instantly breaks apart into water and carbon dioxide (CO_2). It is the carbon dioxide that creates the fizz as it leaves the solution.

MORE: For an extra cool effect, use a tall container, and construct a realistic looking volcano around it using cardboard or paper. Add red food coloring to the vinegar before you add it to the baking soda. This will give it the appearance of lava and will make the "eruption" look even better!

2) DRY ICE BUBBLE

This experiment is a great one for adults and kids to do together. Have fun making a dry ice bubble that fills with fog. How big will it get before it bursts?

Materials you'll need:

- Water
- A bowl with a lip around the top
- A strip of cloth or material
- Soapy mixture (just add a little dishwashing liquid to some water)
- A few pieces of dry ice
- Remember, Safety first! Dry ice as it can cause skin damage if not used safely. An adult should handle dry ice with gloves, and both of you should avoid directly breathing in the vapor.

What to do:

1. Place your dry ice in the bowl and add some plain water (it should start to fog a little bit).
2. Soak the strip of cloth in your soapy mixture. Run it around the lip of the bowl before pulling it straight and dragging it across the top of the bowl. This should form a bubble layer over the dry ice.
3. Stand back and watch what happens!

What's happening?

Dry ice is carbon dioxide (CO_2) in its solid form. At most temperatures, dry ice changes directly from a solid to a gas, without ever being a liquid (a process called *sublimation*). When dry ice is put in water it quickens the process of sublimation, which creates clouds of fog that fill up your dry ice bubble. This continues until the pressure is too much and the bubble explodes, releasing all the built-up fog.

3) LEMON FIZZ

Make your own lemonade soft drink with this fun experiment for kids.

Materials you'll need:

- 1 -2 lemons
- Drinking glass
- Water
- 1 teaspoon of baking soda
- Some sugar

What to do:

1. Squeeze as much of the juice from the lemons as you can into the glass
2. Pour in equal amounts of water
3. Stir in the teaspoon of baking soda.
4. Give it a taste.
5. If you want it to be sweeter, add a little bit of sugar and taste again.

What's happening?

The mixture you created should become bubbly and taste like a fizzy lemonade. These are the same bubbles you'll find in soda and other fizzy drinks. When you add baking soda to the lemon-water mixture, the bubbles that form are carbon dioxide (CO_2). Lemon is an acid, and baking soda

is a base, and combining these two make a chemical
reactions—the fizz!

4) DIET COKE & MENTOS EXPLOSION

The Diet Coke & Mentos Explosion experiment is one of the most popular science experiments you can do at home.

Materials you'll need:

- Large bottle of Diet Coke
- Half a pack of Mentos (any flavor)
- Optional: Funnel or Geyser tube

What to do:

1. Make sure you are doing this experiment outside and in a place where it is okay to make a mess.
2. Stand the Diet Coke bottle upright and unscrew the lid.
3. If you have a funnel, place it on top. If not, skip this step.
4. Drop in all the Mentos (half the pack should be enough) into the Diet Coke and stand way back!

What's happening?

There are a few different theories about this experiment, but most people agree that the combination of carbon dioxide (CO_2) in the Diet Coke and the little dimples on Mentos candy have a huge reaction.

Remember the lemon fizz experiment? The same thing that makes that fizz is what makes soda fizz—Carbon Dioxide (CO_2). When it is added to bottles of soda at a factory, the lid is sealed, and the carbon dioxide doesn't get released until you open it. This means in every bottle of soda, there is a lot of carbon dioxide gas just waiting to escape in the form of fizzy bubbles.

By dropping Mentos into the Diet Coke, you speed up the process. The Mentos breaks the surface tension of the soda and allows bubbles to form on the surface of the Mentos. Because Mentos candy pieces are covered in tiny dimples, its surface area is much larger, allowing for a HUGE amount of bubbles to form.

Why Diet Coke? It has slightly different ingredients which have shown that it is the soda of choice for this particular experiment, but you can use whatever soda you like.

5) BALLOON BLOW-UP

Chemical reactions make for some of the best science experiments. Play with carbon dioxide (CO_2) by funneling it through a plastic bottle to blow up a balloon.

Materials you'll need:

- Balloon
- ¼ cup water, or less
- Plastic water or soda bottle
- Drinking straw
- Juice from one lemon
- 1 teaspoon of baking soda

What to do:

1. First, loosen up the balloon by stretching it, this will make it easier to inflate.
2. Pour the water into your plastic bottle.
3. Add 1 teaspoon of baking soda and stir with the straw until it has completely dissolved.
4. Pour in the lemon juice, and quickly place the balloon over the mouth of the bottle.
5. Sit back and observe.

What's happening?

If all goes according to plan, your balloon should inflate. Adding the lemon juice (an acid) to the baking soda (a base) creates a chemical reaction, resulting in carbon dioxide (CO$_2$). This gas needs a lot of room to spread out, so once it fills the bottle, it moves up and into your balloon! It's the carbon dioxide that pushes the balloon outwards and blows it up.

6) LAVA LAMP!

Have you ever seen a lava lamp before? Maybe your parents had one? Well, not you can make your own with this fun science experiment that teaches you about chemical reactions.

Materials you'll need:

- Water
- A clear plastic bottle
- Vegetable oil
- Food coloring (any color you want!)
- Alka-Seltzer (or other tablets that fizz)

What to do:

1. Pour water into the plastic bottle until it is around ¼ full (if you have a funnel, use it! This will make pouring without spilling much easier).
2. Now pour in the vegetable oil until the bottle is nearly full.
3. Wait until the oil and water have separated, then add 10 drops of food coloring.
4. Watch as the food coloring slowly falls through the oil and mixes with the water.
5. Cut or break an Alka-Seltzer tablet into a few pieces.
6. Drop one piece into the bottle and watch it bubble!
7. When the bubbling stops, add another piece of Alka-Seltzer and repeat!

What's happening?

If you've tried the **Oil & Water Experiment,** then you know that oil and water don't mix very well. When they're together in the bottle, the oil rests on top on top of the water because it has a lower density. When you drop a piece of Alka-Seltzer in, it sank to the bottom, started dissolving, and released small bubbles of carbon dioxide gas (CO_2). As the gas bubbles rose to the top, they took some of the dyed water with them. The gas escapes from the top and the dyed water falls back down to the bottom. Alka-Seltzer (and other fizzy tablets like it) fizzes because it is made of *citric acid* and *baking soda* (sodium bicarbonate). Those two ingredients react with water to form the carbon dioxide (CO_2) gas, which are the bubbles that you see.

You can store your "lava lamp" by simply screwing the cap back on. Anytime you want to play with it or show someone, just add another piece of Alka-Seltzer.

7) EXPLODING SANDWICH BAG

This fun experiment is messy, so be sure to do it outside. It also uses warm water, so be sure to ask an adult for help.

Materials you'll need:

- One small zip-lock bag
- 3 teaspoons of baking soda
- ¼ cup of warm water (ask for help!)
- ½ cup vinegar
- Measuring cup
- 1 piece of tissue

What to do:

1. Outside, pour the warm water into the bag.
2. Add the vinegar to the bag.
3. Place the baking soda in the middle of the tissue, and wrap it up.
4. Zip the bag mostly closed, but leave enough space to add the tissue.
5. Add the tissue to the bag and quickly zip it closed.
6. Put the bag on the ground and back away.

What's happening?

Inside the bag, the baking soda and the vinegar mix to create and *acid-base reaction.* Those two chemicals create carbon dioxide gas (CO_2), and fills up the bag until the bag can no longer hold it in and it should explode!

MORE: Try different temperatures of water, different amounts of baking soda, and different sized bags to see how they affect the reaction.

8) FILM CANISTER ROCKET

This experiment is a lesson in chemistry and air pressure.

Materials you'll need:

- A plastic bottle with a snap top (like an old film canister).
- Water
- Alka-Seltzer tablets
- Optional: Pieces of cardboard and tape to make fins and a nose for your rocket.

What to do:

1. If you have cardboard, cut out small fins and a nose cone for your rocket. Attach it to your bottle with tape. Be sure to attach the nose cone on the bottom of the bottle.
2. Fill up the bottle with water and leave a little bit of space at the top (about ½ an inch).
3. Put in one Alka-Seltzer and quickly snap the bottle closed, and stand back!
4. If one tablet wasn't enough to make it explode, try it again with more.

What's happening?

The explosion happens because Alka-Seltzer is made of citric acid and sodium bicarbonate (baking soda), which acts as a base. When the tablets are added to water they react to form carbon dioxide (CO_2). Carbon dioxide spreads out and creates pressure, and when the pressure builds the top of the container pops off and its contents explode out.

9) SWIMMING SPAGHETTI

Make spaghetti do tricks with this fun, easy, and FIZZY experiment!

Materials you'll need:

- uncooked spaghetti
- 1 cup of water
- 2 teaspoons of baking soda
- 5 teaspoons of vinegar
- tall clear glass

What to do:

1. Put water and baking soda in the glass.
2. Stir until the baking soda is dissolved.
3. Break spaghetti into 1-inch pieces and put 6 pieces in the glass. They will sink to the bottom.
4. Add vinegar to the mixture in the glass.
5. Watch what happens to the pieces of spaghetti.
6. Add more vinegar as the action starts to slow down.

What's happening?

When baking soda and vinegar are mixed together, a chemical reaction occurs, and carbon dioxide (CO_2) gas is produced. This forms a lot of bubbles on top of the mixture and smaller bubbles at the bottom of the glass. These little bubbles stick to the spaghetti and make it float to the surface. When the spaghetti reaches the surface, the bubbles pop and the spaghetti sinks to the bottom.

10) DANCING RAISINS

Use carbon dioxide to make raisins dance!

Materials you'll need:

- 1 can of clear soda (like 7-Up or Sprite)
- Tall, clear drinking class
- Several fresh raisins

What to do:

1. Empty the can of soda into the tall glass (notice the bubbles coming up from the bottom of the glass).
2. Drop 6 or 7 raisins into the glass. Watch the raisins for a few seconds. Do they sink or float?
3. Keep watching; what happens in the next several minutes?

What's happening?

The bubbles in the glass is carbon dioxide gas (CO_2) released from the liquid. Raisins are denser than the liquid in the soda, so initially they sink to the bottom of the glass. The carbonated soft drink releases carbon dioxide bubbles. When these bubbles stick to the rough surface of a raisin, the raisin is lifted because of the increase in buoyancy. Like our **Swimming Spaghetti Experiment,** when the raisin reaches the surface, the bubbles pop, and the carbon dioxide gas escapes into the air. The raisins lose buoyancy and sink back down to the bottom. They keep "dancing" until most of the carbon dioxide has escaped.

11) FIRE EXTINGUISHER

This fun experiment involves the use of matches, candles, and fire, so be sure to ask an adult for help.

Materials:

- Distilled white vinegar
- Baking soda
- Clear glass or container
- A few candles
- Matches

What to do:

1. Pour some baking soda into your clear glass.
2. Pour white vinegar into the same container as the baking soda (watch it start to foam).
3. Wait a few seconds while the reaction calms down.
4. Now, with the help of an adult, ignite the wicks of the candles.
5. Grab your container of solution and pour the air from the container towards the candle's flame. DO NOT pour the liquid!
6. Keep pouring out the air until all the candles are extinguished.

What's happening?

Fire needs *oxygen, fuel,* and *sufficient heat* to stay lit. If you remove any of those 3 components the fire will *extinguish* (go out). In this experiment, our extinguisher is made from the baking soda, which is considered a base, and the vinegar, which is a weak acid. Combining the two makes *carbon dioxide* (CO_2) and water (H_2O). The bubbling is the production of the CO_2 gas. When you pour out the CO_2 gas, it takes away oxygen (one of the 3 things fire needs). The lack of oxygen puts out the flame.

II. SENSE EXPERIMENTS —TEST YOURSELF!

12)TASTE-TESTING EXPERIMENT

Do you know what gives us the ability to taste our food? This easy taste-testing experiment shows that there's a lot more to our sense of taste than you may have thought.

Materials you'll need:

- A small piece of peeled potato
- A small piece of peeled apple (both should be roughly the same size)

What to do:

1. Close your eyes and mix up the potato and apple so you can't tell which is which.
2. Keeping your eyes closed, pinch your nose shut and eat one piece.
3. Keeping your nose closed, eat the other piece.
4. Can you tell the difference?

What's happening?

Pinching your nose shut while tasting the two pieces makes it difficult to tell which is which. This is because our nose and mouth are connected through the same airway. We taste and smell foods at the same time. If you take away your sense of smell, you limit the ability of your brain to tell the difference between certain foods.

13)ARCTIC ANIMAL EXPERIMENT

This fun experiment will teach you how animals in the
Arctic and Antarctic regions keep warm in the cold.

What you'll need:

- Bowl of ice water
- Crisco (or some other vegetable shortening)
- Latex or Vinyl gloves
- Plastic Wrap
- A helper

What to do:

1. First, touch the ice water—if you can place your whole
   hand in it for a second. How does it feel?
2. Next, put on a latex glove (this will help you stay
   clean!)
3. Using your gloved hand, take out a big handful of
   Crisco and make a fist.
4. Have your helper put on gloves and add globs of Crisco
   to the outside of your hand. The goal is to get gloved
   hand completely covered in a big layer of Crisco.
5. Have your helper wrap your Crisco hand with a piece of
   plastic wrap.
6. Now place your hand back into the ice water.
7. How does it feel now?

What's happening?

Animals that live in the icy environments of the Arctic and Antarctic (such as walruses, seals, and polar bears) all have a thick layer of fat, called *blubber*, that keeps them warm in the frigid cold. As you can see, when your hand is covered in fat (the Crisco) your hand stays much warmer than it did without. Now you know how arctic marine animals, like polar bears, stay warm!

14) TEST YOUR DOMINANT SIDE

This experiment will show you how the body and brain work together.

Materials you'll need:

- A pencil or pen
- Some paper
- An empty paper towel tube
- A cup of water
- A small ball

What to do:

1. On the piece of paper, write down LEFT and RIGHT next to each other. As you complete each of the tasks below, make a mark under whichever side you used in the task.
2. When you've finished all the tests, look at your results and draw your own conclusion about which of your sides is dominant.

The Tests:

Eye tests:
Which eye do you use to wink?
Which eye do you use to look through the empty tube?

Hand/Arm tests:
Which hand do you use to write? Can you write with both?

Pick up the glass of water (which hand did you use)?
Throw the ball (which arm did you use)?

Foot/Leg tests:
Run forward and jump off one leg (which leg did you use
to jump)?
Kick the ball (which foot did you use)?

What's happening?

So what side is your *dominant side* (what side do you use
more)? Are you left-handed or right-handed? Do you use
your left foot or your right foot? Do you look more with
your right eye than your left?

MORE: Does your dominant side match your parents?
Have them do the tests with you to see!

15) FLYING MIRROR OPTICAL ILLUSION

This tricky experiment shows you how easily our brained can be fooled. Ask an adult for help with this experiment, plus you'll need another set of eyes to show it to!

Materials you'll need:

- A mirror

What to do:

1. With the help of an adult mount the mirror vertically, on the ground. (try bracing it in between two heave objects to keep it standing up)
2. Now straddle the mirror with one foot in front of the mirror and one in back.
3. Put your weight on the foot on the back side of the mirror and lift the foot in front of the mirror.
4. What does it look like to your observer?

What's happening?

What fools our brain here is the *orientation* of the image. Though the vertical and horizontal orientations of reflected images stay the same, the mirror does change the direction the object is facing into and out of the mirror. Like when you look into a mirror your image looks back at you. So when you keep one leg in front of the mirror, the inside of the leg which would normally face away is now reflected back towards an observer. The observer sees the outside of one leg and the inside of the other. Since the image is in the

same position and orientation as a leg on the other side of
the mirror, our brain is convinced that we're seeing is really
just both your legs. When you lift your leg in front of the
mirror, the image is lifted and our brain is fooled into
thinking that you have lifted both of your legs and are
floating.

16) FILM-CANISTER OPTICAL ILLUSION

Materials you'll need:

- Opaque Film canister
- Pin
- Source of light (such as a lamp)

What to do:

1. Make a small pinhole in the center of the bottom of your film canister.
2. Halfway up the canister, push a pin through the side of the can from the inside, so that the head of the pin is on the inside of the can directly above the pinhole on the bottom.
3. Look into the open end of canister toward a light and adjust the head of the pin until it is in line with the light coming through the pinhole in the bottom.
4. What do you see?

What's happening?

You should see the pin head inside the canister, but it should appear to point in the opposite direction. This is because what you are seeing is the shadow of the pin head, which is right side up on your retina, *superimposed* the bottom of the canister, which is *inverted* on your retina. Your brain interprets the messages from your retina by turning images upside down, which makes the shadow of the pin seem upside down.

17)ARC OPTICAL ILLUSION

Material you'll need:

- 2 identical cardboard coffee sleeves (found at any coffee shop)
- Someone to show the experiment to!

What to do:

1. Unfold and flatten each of the cardboard sleeves.
2. Hold one sleeve above the other with the sleeves arc in the same direction.
3. Make sure that one end is lined up.
4. Ask your audience which one is bigger? They'll say the bottom one!
5. Now switch them. Now the top one is the bottom one, and it look bigger!
6. Place the arcs on top of each other to show your audience that they are actually the same size.

What's happening?

This is an illusion of *comparison*. The segment of a circle seems much longer if it is placed under an identical segment. However both ends need to be lined up on one side. It illusion happens because the longer outside edge of the bottom arc is being compared to the shorter inside edge of the top arc. When the inside curve of the top arc is placed next to the outside curve of the bottom arc, the bottom arc always appears to be bigger.

18)WASHINGTON SMILE

The secret to this optical illusion is all in how you fold the dollar bill.

Materials you'll need
- A 1 dollar bill

What to do:

1. Start with George facing you.
2. At the middle of George's left eye, fold the bill away from you, in half.
3. Do that again in the middle of his right eye. Make those creases are sharp.
4. In between those two folds, fold inward so that the crease is between George's eyes and nose.
5. Pull on the ends of the bill slightly so that you can see his entire face (but making sure that you can still see the folds).
6. Hold George's face in front of you with the face tilted upward. Now slowly begin tilting his face towards the floor.
7. What do you see?

What's happening?

George's smiling face should turn into a frown, but don't worry, it is an optical illusion. When you fold the dollar, your brain and eyes perceive George to show an actual expression, because the folds change the perspective.

19) HOW MUCH CAN YOUR LUNGS HOLD?

How much air can lungs hold? Find out with this fun experiment.

Materials you'll need:

- Clean plastic tubing
- A large plastic bottle
- Water
- Kitchen or bathroom sink

What to do:

1. Plug the bottom of your sink and fill it until it has 10 cm of water in it.
2. Fill the plastic bottle right to the top with water.
3. Put your hand over the top of the bottle and turn it upside down.
4. Place the top of the bottle under the water in the sink and then move away your hand.
5. Push one end of the clean plastic tube into the top of the bottle (keep it under water!)
6. Take a big breath in and put your mouth on the plastic tube.
7. Now, breathe out as much air as you can into the tube.
8. Measure how much the water in the sink has risen and remove the bottle.

What's happening?

As you breathe out through the tube, the air from your lungs pushes out the water from the bottle. The amount of water (the *volume)* you pushed is equal to how much air your lungs can hold. If you have a big air capacity in your lungs, it means you distribute oxygen around your body more quickly.

20)LUNG SIMULATOR

This lung simulator machine uses some complex tools, so be sure to ask an adult for assistance.

Materials you'll need:

- Drill (ask an adult for help!)
- Drinking straw
- 2 rubber bands
- Scissors
- Utility knife
- 2 balloons
- Modeling clay
- Plastic bottle with cap

What to do:

1. With the help of an adult, use the utility knife to cut off the bottom of the bottle.
2. Cut off the tip of one of the balloons, and stretch it over the bottom of the bottle
3. Secure the balloon with a rubber band.
4. Insert the straw into the other balloon and secure it with another rubber band.
5. With the help of an adult, drill a hole in the center of the bottle cap.
6. Put the balloon-covered end of the straw into the bottle, push the straw through the cap, and screw the cap back on.

7. Use the modeling clay to create a better seal around the straw where it exits the bottle cap.
8. Now, pull on the end of the balloon that's wrapped around the bottom of the bottle.
9. What happens?

What's happening?

The balloon inside the bottle should inflate! This happens because you created an *airtight* container, which means air cannot go in or out. When you pull the outer balloon, you increase the volume of the entire container, but the volume of air inside the container remains the same. In this way, you create an area of *low air pressure* inside of the bottle, while the air pressure outside the bottle remains the same. The higher air pressure from outside want to go to the area of low air pressure, so it forces itself into the straw, and into the balloon to balance the air pressure.

21)MAKE BALLOON SPEAKERS

Learn how small sounds can make a big noise with a good sound conductor.

Materials you'll need:

- Balloon
- Your ears

What to do:

1. Blow up the balloon.
2. Tap lightly on the balloon and listen.
3. Now, hold the balloon close to your ear.
4. Now tap lightly on the other side on the other side of the balloon and listen!

What's happening?

Even though you are only tapping lightly on the balloon, your ears can hear the noise loudly, right? By blowing up the balloon, you force the air molecules inside of it to get closer together. Because the balloon's air molecules are closer together, they become a better *conductor* (mover) of sound waves than just the air around you.

22)WHIRLING HOSE

Materials you'll need:

- A piece of irrigation tubing 3 feet long, and 2 inches in diameter (easily found in any hardware store)
- Plastic bag (garbage bag or shopping bag works fine)
- Rubber band or tape

What to do:

1. Hold one end of the tube and twirl the other end over your head in a circle.
2. Spin the tube faster and observe how the sound changes. (Fast twirling creates a higher pitch)
3. Attach the plastic bag to the end of how with the rubber band (or tape).
4. With your mouth close to the opening of the hose, blow! (The bag should inflate with a few big breaths)
5. Once inflated, twirl the house and bag.
6. Listen!

What's happening?

As you twirl the tube, *air molecules* are launched out of the other end. The faster the twirl, the faster the molecules fly out. The plastic bag allows you to see the movement of the air molecules as it deflates and sound is made.

23) CHIMING COAT HANGERS

Materials you'll need:

- Wire coat hanger
- Your ears
- 3-foot long piece of thread
- A rigid object (like a wall)

What to do:

1. Tie the middle of your thread to the hook of your hanger, so you have two 1 ½ foot long pieces of thread on either side.
2. Wrap the ends of the thread around the second middle finger of each hand and stick these fingers in your ears.
3. Stand next to a wall or other rigid object and allow the hanger to bump against it.
4. What do you hear?

What's happening?

When you bump the coat hanger against something, it vibrates which makes a very quiet sound. However it sounds louder when your fingers are in your ears because the vibrations travel better through the tight thread than through the air.

MORE: Try hitting the hangar against other object. Does it change the sound? Try bending the coat hanger. Does it change the sounds?

24)"CHICKEN" SPEAKERS

You might not have a chicken, but with this fun science experiment, you can make your own chicken sounds!

Materials you'll need:

- A plastic drinking cup
- Yarn or cotton string
- 1 paper clip
- Paper towel
- A nail
- Scissors
- Water

What to do:

1. Cut a piece of yarn about 20 inches long.
2. Use the nail to carefully punch a hole in the center of the bottom of the cup (ask for help!)
3. Tie one end of the yarn to the middle of the paper clip.
4. Push the other end of the yarn through the hole in the cup and pull it through as shown in the picture.
5. Get a piece of paper towel about the size of a dollar bill, then fold it once and get it damp in the water.
6. Hold the cup firmly in one hand, and wrap the damp paper towel around the string near the cup. While you squeeze the string, pull down in short jerks so that the paper towel tightly slides along the string. What do you hear?

What's happening?

This is an example of how a *sounding board* works. The vibrations from the string would be almost silent without the cup, but when you add the cup, it spreads the vibrations and makes them louder.

25)MAKE A DUCK CALL

Have you seen the popular TV show Duck Dynasty? They made their fortune making duck calls, now you can make your own!

Materials you'll need:
- One plastic straw
- Scissors

What to do:

1. Use your fingers to press on one end of the straw to flatten it. Try to make it as flat as possible.
2. Cut the flattened end of the straw into a point
3. Flatten it out again real good.
4. Take a deep breath, put the pointed end of the straw in your mouth, and blow hard into the straw. If all goes well you should hear a somewhat silly sound coming from the straw.
5. Try cutting the straw different sizes to see how the sound changes/

What's happening?

All sounds come from vibrations. That little triangle that you cut in the straw forced the two pieces of the point to vibrate very quickly against each other when you blew through the straw. Those vibrations from your breath going through the straw created that strange duck-like sound that you heard.

26)WATER WHISTLE

Learn about the science of sound with this fun water whistle experiment!

Materials you'll need:

- Drinking straw
- Scissors
- Drinking glass
- Water

What to do:

1. 1/3 of the way down the straw, PARTIALLY cut through the straw—you should leave a small piece uncut so that the straw stays together.
2. At the cut mark, bend the straw into a right angle, but be careful not to break it in two.
3. Fill your glass ¾ full with water.
4. Slide the longer side of the straw into the water.
5. Keep the straw at a 90 degree angle, and place your lips on the shorter end of the straw.
6. Lightly blow with a constant breath. What do you hear?

What's happening?

Your whistle makes a sound because of *vibration*. All sounds are *sound waves*. These sound waves are vibrations traveling through the air until they reach your ears. The whistling sound in this experiment is caused by vibration of air inside the straw. When you blow the air across the top of the straw segment in the water, you are causing the column of air inside to vibrate.

MORE: Try raising or lowering the straw in the water. What happens to the pitch of your whistle when you do this?

27) SINGING WINE GLASS

Learn how to make a wine glass sing in this experiment. Ask an adult for help with the glass.

Materials you'll need:

- A wine glass
- Water
- Napkin

What to do:

1. Fill the glass half-full with water.
2. Dip your pointer finger into the water to clean it.
3. Use a napkin to wipe clean your finger (the cleaner the better).
4. Dip your finger into the water again to moisten it.
5. Set your clean, damp finger on the rim of the glass, press down slightly, and rub it all the way around the rim without stopping.
6. Keep going in a circular motion along the lip of the glass while maintaining the pressure.
7. What do you hear?

What's happening?

As you rub your finger on the rim of the wine glass, your finger slightly sticks to the glass and then slides. This happens in very short lengths and makes a vibration inside the glass which in turn produces a sound. Having a clean finger improves stick-and-slide action and subsequent vibration. Once you get started, the glass *resonates*, which means you're causing the crystals in the glass to vibrate together and create one clear tone.

MORE: You can change the pitch by adding or taking away water in the glass.

28) BUZZ MAKER

Create a simple noise maker in this experiment.

Materials you'll need:

- Index card
- Wide rubber band
- 2 foam pieces with adhesive backing
- String (yarn works too)
- Jumbo craft stick
- Scissors

What to do:

1. Using your scissors, cut two of the corners off one of the long sides on the index card.
2. Cut two pieces of adhesive foam down to 2cm x 5cm.
3. Place the jumbo craft stick on the long, uncut side of the index card (only half of the stick should be touching the index card)
4. Fold one piece of the adhesive foam around the end of the craft stick and index card to hold them in place.
5. Now cut your string to be 3 feet in length.
6. Place the string across the other piece of adhesive foam (with most of the string on one side and leave 5-10 cm on the other).
7. Fold that piece of foam around the other end of the craft stick and index card. Tie a loop around the foam with the 5-10 cm string part.

8. Stretch a wide rubber band over each of the foam pieces (make sure the rubber band isn't twisted)
9. Now take the long end of the string and twirl the Buzz Maker over your head.

What's happening?

The buzzing noise is produced by the rubber band *vibrating* against the craft stick. The vibrations are caused by air moving around the rubber bands. The speed of your twirl directly affects the pitch of the noise it makes. The faster you spin your Buzz Maker, the higher the pitch will be.

MORE: How does the noise change when you shorten or lengthen the string? What happens when you change the twirling speed of the Buzz Maker?

29) SCREAMING BALLOON

Materials you'll need:

- A clear, latex balloon
- A ¼" hex nut (easy to find in any hardware store)

What to do:

1. Drop the hex nut into the balloon and make sure it goes all the way in.
2. Blow up the balloon (but be careful not to overinflate it), and tie it off.
3. Grasp the balloon with an open hand at the stem end. The neck of the balloon should be in your palm and your fingers and thumb will grip the balloon down the sides.
4. While holding the balloon, palm down, turn it in a circular motion. The goal is to get the hex nut to roll around inside the balloon.
5. What do you hear?

What's happening?

This experiment shows how sounds can come from *friction*, which occurs when objects rub against each other. In this case, the *friction* is the 6 sides of the hex nut rubbing against the inside of the balloon. The screaming sound comes from the vibration that occurs against the inside wall of the balloon.

III. DENSITY SCIENCE EXPERIMENTS

30)LAVA IN A CUP

This experiment like the **Lava Lamp** project, but using different ingredients to learn about a different reaction.

Materials you'll need:

- A clear drinking glass
- ¼ cup vegetable oil
- 1 teaspoon salt
- Water
- Food coloring

What to do:

1. Fill the drinking glass with water until it is ¾ full.
2. Add about 5 drops of food coloring, whichever color you like.
3. Slowly pour the vegetable oil into the glass until it is full.
4. Now sprinkle the salt on top of the oil.
5. Watch what happens!

What's happening?

Like the first **Lava Lamp** experiment, the oil floats on top because it less dense than water. When you add salt to the mix, it sinks down into the water (but takes some oil with it). Once it gets to the water, the salt dissolves, but the oil goes back up, taking some colorful water with it. If you want to see the cool trick again, just add more salt!

31)OIL & WATER EXPERIMENT

Some things just don't get along well with each other. Take oil and water as an example, you can mix them together and shake as hard as you like but they'll never become friends.....or will they?

Materials you'll need:

- Plastic bottle with lid (an empty water bottle or soda bottle will work perfectly)
- Water
- Food coloring
- 2 Tablespoons of cooking oil (any kind will do)
- Optional: Dish washing liquid or detergent

What to do:

1. Add a few drops of food coloring to the water.
2. Pour 2 Tablespoons of the water along with 2 Tablespoons of cooking oil into the plastic bottle.
3. Screw the lid on tight and shake the bottle as hard as you can.
4. Put the bottle back down watch.

What's happening?

It may have seemed like the oil and water were mixing together, but the oil will float back to the top eventually. While water can mix with most other liquids, it cannot mix with oil. Because water molecules are strongly attracted their own molecules, and the same goes for oil, they separate into themselves and don't mix together. The oil floats above the water because it has a lower density.

MORE: Try adding a little dish washing liquid or detergent. Soap is attracted to both water and oil, which helps them all mix together. This is called an *emulsion*. This is how detergent helps clean greasy dishes—the soap takes the oil and grime off the plates, into the water, and down the drain.

32) SINK OR SWIM: ORANGE EXPERIMENT

Do you know if an orange sinks or swims when placed in water? This fun experiment will teach you about density while also revealing a special characteristic about oranges.

Materials you'll need:
- An orange
- A deep container (or large mixing bowl)
- Water

What to do:

1. Fill your large container with water.
2. Put the orange in the water and watch for a few seconds.
3. Next, peel the rind from the orange and try the experiment again.
4. What happens now?

What's happening?

The first time you put the orange in the water, it probably "swam" on the surface. However, once you removed the rind, it sunk to the bottom.

This is because an orange rind is full of tiny air pockets, which gives it a lower *density* (mass and volume) than water. Its lower density lets it float on the surface of the water. By removing the rind (and all the air pockets) its density is higher than the water, which makes it sink.

33) SINK OR SWIM: SODA CANS

Materials you'll need:

- several unopened cans of regular soda
- several unopened cans of diet soda
- a large ice chest or sink

What to do:

1. Fill the ice chest or sink almost to the top with water.
2. Place a can of regular soda into the water (be sure no air bubbles are trapped under the can when you place it in the water). What happens, does it sink or swim (float)?
3. Now try it with a can of diet soda. Does it sink or float?

What's happening?

Even though the cans are the same size, with the same amount of soda in them, their *density* is different due to what is dissolved in the soda (their ingredients). Regular soda contains sugar A LOT of sugar. Some contain around 40 grams of sugar. Diet sodas use artificial sweeteners. These artificial sweeteners are much sweeter than regular sugar, so only a few grams of artificial sweetener is in diet soda. The difference in the amount of dissolved sweeteners leads to a difference in density. Cans of regular soda are *denser* than water, so they sink, whereas, cans of diet soda are usually less dense than water, so they swim.

34)MAGIC KETCHUP EXPERIMENT

You can make a pack of ketchup float and sink at your command….No, it's not really magic—it's science!!

Materials you'll need:

* A 1 liter plastic bottle
* Ketchup pack from a fast food restaurant
* Salt

What to do:

1. Add a ketchup pack to the bottle.
2. For the floating ketchup pack, just screw the cap on the bottle and squeeze the sides of the bottle hard. If the ketchup sinks when you squeeze it, and floats when you release it, you're ready to show it off.
3. If the ketchup pack sinks, add about 3 tablespoons of salt to the bottle.
4. Cap it and shake it up until the salt dissolves.
5. Continue adding salt, a few tablespoons at a time until the ketchup is just barely floating to the top of the bottle, then cap it tightly.
6. Now squeeze the bottle. The ketchup packet should sink when you squeeze the bottle and float up when you release it.

What's happening?

This experiment is all about *buoyancy* and *density.*
Buoyancy describes whether objects float or sink. This

usually describes how things float in liquids, but it can also describe how things float or sink in and various gasses. Adding salt to the water adjusted the water's density to get the ketchup to float. When you squeeze the bottle hard enough, you put pressure on the packet. That causes the bubble to get smaller and the entire packet to become denser than the water around it and the packet sinks. When you release the pressure, the bubble expands, making the packet less dense (and more buoyant) and it floats back up.

35)CREATE A CARTESIAN DIVER

This experiment was invented by philosopher Rene Descartes in order to demonstrate the principle of buoyancy.

Materials you'll need:

- A clear 1 liter plastic soda bottle with cap
- A ball point pen cap that does without holes on the top
- Some modeling clay

What to do:

1. Remove any labels from your bottle so that you can watch the action.
2. Fill the bottle to the very top with water.
3. Place a small amount of modeling clay at the end of the point on the pen cap.
4. Slowly place the pen cap into the bottle, modeling clay end first. It should just barely float. If it sinks take some clay away. If it floats too much add more clay.
5. Now screw on the bottle cap nice and tight.
6. You can make the pen cap rise and fall at your command.

What's happening?

This experiment is like our **Magic Ketchup Experiment—**it's all about density. When you squeeze the bottle, the air bubble in the pen cap compresses and that makes it denser than the water around it. When this happens, the pen sinks.

When you stop squeezing, the bubble gets bigger again, the water is forced out of the cap, and the pen cap rises.

36) SALT & MARBLE EXPERIMENT

Materials you'll need:

- Plastic test tube a cap (or cork)
- Salt
- Marble

What to do:

1. Fill the test tube ¾ full of salt.
2. Place your marble on top of the salt.
3. Close the end of the test tube with a cap
4. Now try to get the marble from one side of the tube to the other, through the salt.
5. What happens?

What's happening?

You may think that because the marble is *denser* than the salt, it will sink to the bottom when salt is agitated. However, the opposite is true! To make it work, hold the tube vertically so the marble is near the bottom. As you shake the tube up and down, the marble will rise through the salt. Each time the tube is jolted upwards, both the marble and the salt move up at the same speed. Because the salt particles are lighter and smaller, they experience more *relative friction* than the marble, which makes the salt particles slow down more quickly. Each time you shake the tube, more salt is packed underneath the marble, until it eventually it emerges from beneath salt.

37) FIREWORKS IN A JAR

This colorful experiment uses warm water, so be sure to get the help of an adult!

Materials you'll need:
- Warm Water (ask for help!)
- 2 Tablespoons of oil
- Food coloring
- Tall jar
- Cup
- Fork

What to do:

1. With the help of an adult, fill the jar with warm water
2. Put the two spoons of oil into your cup, and add a few drops of food coloring
3. Use the fork to break up the drops of food coloring into smaller drops.
4. Pour the oil mixture into the jar of water and watch!

What's happening?

The food coloring should slowly sink out of the oil into the water, expanding as they fall. Food coloring dissolves in water, but not in oil. The colored drop will begin to sink because they are *denser* (heavier) than the oil. Once they sink into the water, they start to dissolve, which makes them look like tiny, colorful fireworks!

IV. VINEGAR AND ACID EXPERIMENTS

38) SHINY PENNIES

Do you have any old, dirty pennies? Pennies so dirty you can't even see Abraham Lincoln or the date? This experiment will show you how to use chemistry to clean them!

Materials you'll need:

- A few old (not shiny) pennies
- 1/4 cup white vinegar
- 1 teaspoon salt
- Non-metal bowl
- Paper towels

What to do:

1. Pour the vinegar into the bowl and add the salt.
2. Stir them together.
3. Put about 5 pennies into the bowl and count to 10/
4. Take out the pennies and rinse them out in some water.

What's happening?

There is some pretty fancy chemistry going on in that little bowl of yours. It turns out that vinegar is an acid, and the acid in the vinegar reacts with the salt to remove what

chemists call *copper oxide* which was making your pennies
dull.

39) BEND A BONE!

Next time your family has chicken for dinner, ask one of your parent's to save you a chicken bone for this awesome vinegar experiment!

Materials you'll need:

- A large glass jar
- A chicken bone
- Vinegar

What to do:

1. Rinse off the chicken bone by running it under water (make sure all the meat is off)
2. Feel the bone, and observe how hard it is. Try to bend it.
3. Now place the bone into the jar and cover it with vinegar. Put the lid back on.
4. Leave it for at least 3 days, and then remove it.
5. Rinse it off and try bending it again. Feels different doesn't it?

What's happening?

Vinegar is considered a weak *acid,* but it is still strong enough to dissolve away the *calcium* in the chicken bone. Calcium is what keeps bones hard, so once it's dissolved, all that's left is soft bone tissue, which you can bend!

40) STEEL & VINEGAR CHEMICAL REACTION

This is a great science experiment for learning about exothermic chemical reactions.

Materials you'll need:

- Steel Wool
- Vinegar
- Two tall drinking glasses
- A piece of paper
- Thermometer

What to do:

1. Place the steel wool in one of the glasses.
2. Pour vinegar into the glass, over the steel wool, and allow it to soak.
3. After soaking for one minute, remove the steel wool and let excess vinegar drain out.
4. Wrap the steel wool around the bottom of the thermometer and put them both in the second glass.
5. Cover the glass with paper or cardboard to keep the heat in. (If your thermometer is taller than the glass, make a hole in the paper so the thermometer to go through).
6. Check the temperature at the very beginning and write it down.
7. Watch for five minutes and record your results.

What's happening?

The temperature inside the glass should gradually rise (it may even get a little foggy inside). This is because soaking the steel wool in vinegar removes the protective coating and allows the iron (in the steel) to rust. Rusting (a process called *oxidation*) is a chemical reaction that happens between iron and oxygen. This reaction creates heat, which increases the temperature around it. When a chemical reaction releases energy in the form of heat, it is called an *exothermic reaction.*

41)ICE TRAY BATTERY

Use this experiment to turn some everyday items—like an ice tray—into a working battery!

Material you'll need:

- Distilled white vinegar
- 5 pieces of copper wire
- 5 galvanized nails
- Ice tray
- 1 LED light (with a two leg plug)

What to do:

1. Wrap one nail with a piece of copper wire, and leave some of the wire extended from below the head of the nail.
2. Do the same with the rest of the nails and copper wire so you have 5 pieces total.
3. Pour distilled white vinegar into 6 wells of your ice tray.
4. Insert each nail into a well of vinegar, while placing the extended wire into the next well. This will leave you with 1 vinegar well with just a copper wire inside, as well as one 1 vinegar well with just a nail in it.
5. Take your LED light, and place on of its legs into the well with just a copper wire inside it and place the other leg into the well with just a nail in it.
6. The bulb should light up! If not, flip the legs around.

What's happening?

A *Voltaic Battery* is made of two different metals suspended in an acidic solution. In this experiment, the two metals are *zinc* and *copper*. The copper is in the copper wire, and the zinc is in the nail. The acidic solution is the *acetic acid* in the distilled white vinegar.

The two metals work as *electrodes,* which is where electrical current enters and leaves the battery. With the zinc and copper arrangement here, the electrical current flows out of the wire and into the nail, while passing through the acidic solution. Once it is connected to the LED light, the circuit is complete and it can power on the light.

V. INCREDIBLE EGG EXPERIMENTS

42)EGG FLOTATION

Use this egg experiment to learn about density. Normally, an egg will sink to the bottom of a glass of drinking water, but what will happen if you add salt?

Materials you'll need:

- One egg
- Water
- Salt
- A tall drinking glass

What to do:

1. Pour water into the glass until it is half full
2. Stir in 6 big spoonful's of salt
3. Add more water until the glass is almost full (pour carefully and be careful not to mix the contents)
4. Gently lower the egg into the water

What's happening?

The salty water should be dense enough for the egg to float in the middle of the glass. This happens because the denser the liquid, the easier it is for your egg (or any object) to float in it. Salt water is denser than plain old tap water.

When you lower the egg into the glass, it drops through the plain tap water until it reaches the salty part.

43) EGG TESTING

Can you tell the difference between a raw and hardboiled egg? They look the same, but which is which? Answer that tricky question and learn about inertia with this fun experiment.

Materials you'll need:

- Two eggs, one raw and one hard-boiled. (Refrigerated long enough to be the same temperature).

What to do:

1. Spin the eggs and watch!
2. One egg should spin while the other wobbles.
3. Spin again and try touching each of the eggs while they are spinning.
4. One egg should stop quickly, and the other should continue to move.

What's happening?

Which one wobbles? It should be the raw egg, because the raw white and yolk move around inside the shell, which causes it to wobble. Its center of gravity (its insides) continues to move due to *inertia*, the same type of force you feel when you stop suddenly in a car, your body wants to continue moving while the car wants to stay in the same place. With the second part of the experiment, *inertia* also causes the raw egg to continue spin even after you have touched it, while the solid insides of the hard-boiled egg causes it to respond more quickly your touch.

44) RUBBER EGG

It's easy to make a rubber egg if you understand chemistry. Learn how with this easy experiment!

Materials you'll need:

- Raw egg
- Tall glass
- Vinegar

What to do:

1. Place the egg the glass.
2. Pour vinegar over it until it is completely covered.
3. Leave the egg in the vinegar for at least 24 hours.
4. On the second day, pour out the vinegar and replace it.
5. Now place the glass in a safe place for a week. Don't disturb it, but play close attention to what is happening to the bubbles forming on the surface of the shell.
6. Seven days later, pour out the vinegar and rinse the egg with water. The egg should look translucent!

What's happening?

Vinegar is an acid. Egg shells are *calcium carbonate.* The vinegar reacts with the calcium carbonate and breaks it down into separate part so calcium and carbonate. The calcium floats around and the carbonate reacts to form the carbon dioxide (CO_2) bubbles that you see all around the shell.

45)CHIMNEY EGG

This experiment tests the ideas that you can get an egg into a bottle, simply by changing the air pressure! This experiment uses fire, so be sure to ask an adult for help.

Materials you'll need:

- Two hard boiled eggs
- A bottle with a large mouth
- Lighter or matches (ask for help!)
- Paper

What to do:

1. Be sure you are using hard-boiled eggs.
2. Now, have an adult light your piece of paper on fire.
3. Quickly drop the flaming paper into bottle.
4. Place the egg top of the bottle.
5. Wait and see what happens!

What's happening?

When you add the hot flame inside the bottle, you cause the air molecules to move around faster and farther away from each other. This causes the air to expand, and with nowhere to go (because the egg blocks the top), the air pressure increases, a process called *expansion.* When this happens, the egg is pushed out of the way a little, allowing some of those molecules to escape. When the air cools down, the molecules move a little slower and come closer together, a process called *contraction*.

Now there is a less air in the bottle than there was before. Because of this, the air pressure is now lower on the inside of the bottle, and higher on the outside. This makes the outside air to PUSH on the egg, which pushes it clear the small hole on top and into the bottle!

46)BREATHING EGG

Did you know that an egg shell is covered in little tiny holes (called *pores*)? Prove their existence with this experiment while you make an egg "breathe." This experiment uses hot water, so be sure to get the help of an adult.

Materials you'll need:

- A clear glass or jar
- Hot water (ask for help!)
- An egg
- A magnifying glass

What to do:

1. Carefully place the egg into the glass.
2. Slowly pour hot water into the glass until it is almost full.
3. Leave the glass on a table and watch the egg closely for a few minutes, but don't touch as the glass will be hot.
4. Use your magnifying glass to examine more closely what is happening to the egg.

What's happening?

You should be able to see tiny bubbles forming on the egg shell, that eventually bubble their way to the surface of the water. Every egg has a small air pocket between the shell and the white of the egg. When the air becomes heated (due to the hot water), it expands, and escapes through the egg shell's *pores*. Pores are tiny holes that are too small to see with just your eyes, but when you use a magnifying glass, you can see that egg shells has thousands of them. (By the way, human skin has pores all over it too). The pores allow air to pass through the shell, which makes it look like the egg is breathing as the bubbles escape.

47)EGG WHITE SEPARATION

A simple and scientific experiment to remove the yolk from the white of an egg.

Materials you'll need:

- 1 egg (or more if you want to repeat the experiment)
- Empty plastic water bottle
- Bowl

What to do:

1. Crack an egg into your bowl. Be careful so that you don't break the yolk.
2. In your hand, lightly squeeze an empty plastic water bottle (but don't crush it).
3. Hold the water bottle in the squeezed position and don't release it.
4. Touch the mouth of the water bottle to the egg yolk and slowly release the squeeze on the bottle.
5. What happens?

What's happening?

The egg yolk should be "pulled" into the bottle, leaving the egg white in the bowl. When you squeeze the bottle, you are decreasing the air inside. Releasing the squeeze on the bottle allows air to rush back into the bottle. When you place the mouth of the bottle to the egg yolk, the available *volume* inside the bottle is filled by the yolk. The egg yolk separates from the egg white because of their different *viscosity*. While the egg white is runny and slimy, the yolk is more solid which makes it easier to "suck" up into the bottle.

48)EGG LAUNCH

This incredible egg experiment demonstrates gravity and motion.

Materials you'll need:

- Cardboard tube (toilet paper or paper towel tube works great)
- Pie pan
- Eggs
- Water
- A large drinking glass
- Optional: Tray

What to do:

1. Fill the large drinking glass about ¾ full of water.
2. Place a pie pan on top of the glass. Makes sure it's centered.
3. Place the cardboard tube on the pie plate, positioning it directly over the water.
4. Carefully set the egg on top of the cardboard tube.
5. With your hand, hit the edge of the pie pan HORIZONTALLY. (Don't swing up or down, just hit it solidly from the side (horizontally).
6. What happens?

What's happening?

You applied enough *force* to the pie pan to cause it to fly out from under the cardboard tube. The edge of the pie pan hooked the bottom of the tube, which took the tube with the pan. The force of gravity took over and pulled the egg straight down toward the center of the Earth. The egg was going to keep moving until something stopped it, and luckily you had a glass of water to give a safe place for the egg to stop moving.

49)EGG CRUSH

Materials you'll need:

- Eggs
- sink
- Ring

What to do:

1. Place an egg in the palm of your hand.
2. Close your hand so that your fingers are completely wrapped around the egg.
3. Try as hard as you can to crush it.
4. Now hold the egg between your thumb and pointer finger, and squeeze the top and bottom of the egg.
5. Now, standing over a sink, put a ring a finger of your squeezing hand. Give the egg a squeeze.
6. What happens this time?

What's happening?

The shape on an egg is similar to a *3-dimensional arch,* one of the strongest architectural forms. This means the egg's shape gives it amazing strength. It is strongest at the top and the bottom, which is why the egg doesn't break when you add pressure to both ends. When you completely surround the egg with your hand and squeeze, the curved shape distributes the pressure evenly all over the egg. However, when there is an uneven force of pressure, the egg cannot stand, which is why the egg cracked easily when you wore a ring.

VI.   MAKING GOO, SLIME, & MORE

50) LEARN HOW TO MAKE YOUR OWN SLIME (METHOD 1)

Are you ready to make some fake slime? Part of the fun is grossing out our friends, but this experiment also teaches us about proteins. This experiment requires boiling water, so be sure to get the help of an adult.

Materials you'll need:

- Boiling water (ask an adult for help)
- A cup
- Gelatin
- ¼ cup Corn syrup
- A teaspoon
- A fork

What to do:

1. Pour boiling water into a cup until it is half full.
2. Add three teaspoons of gelatin to the cup.
3. Wait 1-2 minutes for it to soften, and then stir with a fork.
4. Add ¼ cup of corn syrup.
5. Stir the mixture again with your fork.
6. Use your fork to pull out the long strands of gunk that have formed.

7. As the mixture cools, continue to add small amount of
    water to it to keep it "slimy."

What's happening?

This slime, much like the mucus (or snot) in our noses is
made mostly of sugars and protein. Although your
homemade slime is different than the real stuff in your
nose, it too is made of sugars and protein. The long strands
of gunk in your slime are *protein strands*. It is the protein
strands make your slime (and snot) sticky and stretchy.

51) LEARN HOW TO MAKE YOUR OWN SLIME (METHOD 2)

Here's another way that you can make your own slime, but using different ingredients to get a different reaction!

Materials you'll need:

- ¼ cup of water
- ¼ cup of Elmer's glue (or any other craft glue)
- ¼ cup of liquid starch
- Large bowl
- Large spoon
- Optional: Food Coloring

What to do:

1. Pour your glue into the mixing bowl.
2. Add the water to the glue
3. Mix them together.
4. If you have food coloring, add about 6 drops now.
5. Add the liquid starch and stir.
6. Your concoction should be nice and slimy now. The more you play with your slimy mixture, the easier to handle it will become.
7. Have fun playing with your "blob" creation, and just be sure to store it in a plastic bag when you are not using it.

What's happening?

Glue is a *liquid polymer*. This means that the molecules in the glue look like strands in a chain. When you add the liquid starch, it acts as a cross-linker, and binds all of the polymer strands of the glue together. It is the combination of the two that makes your concoction stretchy.

MORE: Here are some additional experiments you can do. Change the amount of water or glue to see if it changes the feel of the slime? Use a different kind of glue to see if it makes better slime? Try storing your slime in different places--what happens to it if it is stored out of a bag instead of inside a bag?

52) PLASTIC MILK

This experiment deals with very hot liquids so get the help of an adult.

Materials you'll need:

- One cup of milk
- 4 teaspoons of white vinegar
- A bowl
- A strainer

What to do:

1. Heat up the milk until it is hot, but not boiling (ask for help!)
2. Now ask the adult to carefully pour the milk into the bowl
3. Add the vinegar to the milk and stir it up with a spoon for about a minute
4. Pour the milk through the strainer into the sink/
5. What should be left in the strainer is a mass of lumpy blobs.
6. When it is cool enough, you can rinse the blobs off in water while you press them together.
7. Now mold it into a shape and play with it for days!

What's happening?

You made a substance called *casein*. This happens when the protein in milk combines with the acid in vinegar. The casein in milk does not mix with the acid and so it forms blobs.

53)MAKE QUICK SAND

Quick Sand is a mesmerizing substance, and now you can make some of your own! Show your friends how it works and they will be amazed.

Materials you'll need:

- 1 cup of corn flour
- ½ cup of water
- Large plastic container
- A spoon

What to do:

1. Thoroughly mix the corn flour and water in your container to make your own "instant" quick sand.
2. Stir slowly to show that it is a liquid.
3. Stir quickly to make it hard, then try to poke it quickly.

What's happening?

Stirring corn flour and water together quickly mixes up the corn flour grains, making it difficult for them to "slide" over each other because there is less water between them. This makes it thick. Stirring slowly allows more water in between the grains of corn flour. This gives it a more liquid texture, as the grains can "slide" over each other more easily.

Poking it quickly has the same effect, making the mixture very hard. If you poke it slowly, the substance doesn't mix in the same way, leaving it runny.

Remember that quick sand can be messy. Play with it outside and don't forget to stir just before you use it.

54)MAKE A BOUNCY BALL

Make your own bouncy ball in this fun experiment!

Materials you'll need:

- Borax
- warm water
- corn starch
- glue
- 2 small mixing cups
- a stirring spoon
- Optional: food coloring

What to do:

1. Label one cup 'Borax Solution' and the other cup 'Ball Mixture'.
2. Pour 4 ounces of warm water into the cup labeled 'Borax Solution' and 1 teaspoon of the borax powder into the cup. Stir the mixture to dissolve the borax.
3. Pour 1 tablespoon of glue into the cup labeled 'Ball Mixture'.
4. Add 3-4 drops of food coloring.
5. Add 1/2 teaspoon of the borax solution you just made and 1 tablespoon of cornstarch to the glue. Do not stir.
6. Allow the ingredients to interact on their own for 10-15 seconds and then stir them together to fully mix.
7. Once the mixture becomes impossible to stir, take it out of the cup and start molding the ball with your hands.

The ball will start out sticky and messy, but will solidify as you knead it.

What's happening?

This activity demonstrates an interesting chemical reaction, primarily between the borax and the glue. The borax acts as a "cross-linker" to the polymer molecules in the glue – basically it creates chains of molecules that stay together when you pick them up. The cornstarch helps to bind the molecules together so that they hold their shape better.

55) DANCING GOO

This dancing goo takes on many different forms—when you play with it quickly, it acts like a solid, when you let it rest, it acts like a liquid.

Materials you'll need:

- 2 cups of Corn Starch
- 1 cup of water
- Subwoofer
- Thin cookie sheet
- Music
- Food Coloring

What to do:

1. Mix together the corn starch and water to form the "goo."
2. Place the cookie sheet on top of the subwoofer, and pour in the goo.
3. Add in a few drops of food coloring (different colors make it more fun!).
4. Hold onto the cookie sheet and play different types of music and different volume levels.
5. Watch it dance!

What's happening?

You are watching your goo react to vibrations. Sound isn't just about what you hear, it is also about vibrations, which become more pronounced on top of the subwoofer. If you added food coloring, you can see how much it moves!

56) MAKE SILLY PUTTY

You don't have to buy Silly Putty, you can make your own concoction at home! Follow the directions in this experiment to make silly putty that you can squish in your hands and mold into shapes like real silly putty.

Materials you'll need:

- 2 containers (one large and one small)
- Water
- Food coloring
- Elmer's glue (or any other kind of PVA glue)
- Borax solution (made my combining 1 Tablespoon of borax with 1 cup of water)

What to do:

1. Cover the bottom of the larger container with your glue.
2. Add a little bit of water and stir.
3. Now add 2-3 drops of food coloring and stir.
4. Add a little bit of your borax solution
5. Stir the mixture up and pour it into the smaller container.
6. How is the mixture looking? It should be joining together and acting like Silly Putty!

What's happening?

The Elmer's glue you used is a type of *polymer* called *polyvinyl acetate* (PVA for short). The borax you used is made of a chemical called *sodium borate*. When you combine the two with water, the borax reacts with the glue molecules and joins them together into one giant molecule. This new mixture is able to absorb large amounts of water, which produces a Silly Putty like substance. You can keep the homemade silly putty, just be sure to store it in a sealed plastic bag when you aren't playing with it.

57) SLIME YOU CAN EAT

All of our goo experiments are *inedible* (meaning you shouldn't eat them!), but this one uses materials that you actually can eat! Gross out friends and family with this cool, slimy experiment. This experiment uses hot materials, so be sure to get some adult help.

Materials you'll need:

- Microwave (get some help!)
- Microwave-safe bowl
- Fiber powder (like Metamucil)
- Measuring cup
- Measuring spoon
- Oven mitt
- Food coloring

What to do:

1. Measure 2 cups of room temperature water into a microwave-safe bowl.
2. Add 2 teaspoons of fiber powder to the bowl of water.
3. Put 2-4 drops of whatever color food dye you like to the bowl. Stir well
4. Heat the bowl in the microwave for 2-4 minutes (or until it starts to boil).
5. With the help of an adult, and using an oven mitt, remove the hot bowl, give it a stir, and then put in back in the microwave for 1-3 more minutes (or until it boils again).

6. Again, with the help of an adult and an oven mitt, take out the mixture, and let it sit until it's cool.
7. How does your slime look?

What's happening?

Fiber powder contains an ingredient called *psyllium hydrophilic mucilloid*. Yeah, it's a long one! When psyllium *hydrophilic mucilloid* is combined with water and heated, they form a slime! You can play with it and eat it!

VII. WATER AND ICE

58)WATER DISPLACEMENT EXPERIMENT

This simple experiment will teach you about water displacement. See how many paper clips it takes to make a full glass of water overflow.

Materials you'll need:
- clear plastic cup
- 100 small metal paper clips

What to do:

1. Fill the cup all the way to the top with tap water.
2. Guess how many paper clips it will take to make the water overflow, and write down your guess.
3. Carefully drop one paper clip at a time into the cup, and count how many it takes until water overflows from the cup. Was your guess close?
4. Look at the top of the water glass from the side, can you see what the water is doing?

What's happening?

The surface of the water should be bulging up over the glass. Drops of water (molecules) stick to each other, which is why the water bulges up when you add the paper clips. This is called *surface tension*.

59) WORMY WRAPPER

Use this simple science experiment to transform a wrapper into a worm. You'll never be bored at a restaurant again!

Materials you'll need:

- Plastic drinking straw, with wrapper
- Water

What to do:

1. To open the straw, carefully open the end of the straw wrapper (rip as little as you can).
2. Gently slide the opened wrapper down the straw. When you pull the straw out, the wrapper should look like an accordion.
3. Stick your straw into a glass of water and put your finger over the hole. Pull out the straw, keeping your finger over the top hole the entire time.
4. Now release your finger in a few times, dropping 1-3 drops onto the crinkled straw wrapper.
5. What does it look like?

What's happening?

All paper (including the straw wrapper) are made up of millions of individual *fibers*. When you crinkle the wrapper, these fibers are bent, crushed, and curled together. By adding water you hydrate the fibers, causing them to swell and thicken. The water continues to rush into the fibers, and they continue to straighten out, creating the appearance of a worm!

60) TRICK HAND BOILER

You can challenge your friends with this experiment—is your body hot enough to make water boil?

Materials you'll need:

- Water
- Small drinking glass
- Piece of cloth
- Rubber band
- Optional: Food coloring
- A friend

What to do:

1. Fill your glass almost to the brim with water. (If you have food coloring, add a few drops now).
2. Cover the top of the glass with the cloth and secure it with a rubber band.
3. Place your hand over the cloth and quickly turn the glass over so that the cloth is facing down and remove your hand. (No water will come out!)
4. With the contraption still facing down, ask a friend to take two fingers and warm them up by rubbing them on their sleeve.
5. (You should be holding the glass with one hand and holding the edges of the cloth with the other.)
6. Ask your friend to hold their fingers beneath the cloth without touching it.

7. While your friend's fingers are below the cloth, you
    should slightly push down on the glass with the hand
    that's holding the glass, while slightly pulling up on the
    cloth with the other.
8. The water inside of the cup will begin to boil!

What's happening?

Even though your piece of cloth isn't waterproof, it seems
to keep the water in just fine in this experiment. Even
though fabric has millions of tiny holes, the water doesn't
come out, because of *surface* tension, which is the
molecules of water bonding to each other. The bubbles
happen because there is air being let into the cup when you
stretch the cloth. Stretching the cloth pulls the holes in the
cloth wider than normal, and allows the air molecules to fill
the little bit of space in the cup, and make bubbles.

61)MAKE WATER GLOW

Make glowing water with the help of a black light in this fun science experiment for kids.

Tonic water doesn't look very strange under normal light but what happens when you look at it under a black light? Does the dye from a highlighter pen do the same thing? Find out what happens and why it happens with this cool experiment that you can do at home.

Materials you'll need:

- A black light
- Tonic water
- A highlighter pen.
- A dark room

What to do:

1. Take the Tonic water with you into a dark room.
2. Turn on the black light and hold it near the Tonic water, how does it look?
3. Next, if you are using a highlighter pen, carefully break it open, remove the felt and soak it in a small amount of plain tap water for a few minutes.
4. Return to the dark room with your glass of water and highlighter felt.
5. Turn on the black light again, how does it look?

What's happening?

The black light creates *ultra violet (UV) light*, which illuminates things called *phosphors*. Phosphors turn UV light (light we can't see) into visible light. Both the tonic water and the dye from highlighter contain phosphors, which is why they glow in the dark when you shine a black light on it.

62)LEAK-PROOF MAGIC

This science experiment will teach you more about polymer chains while keeping you dry. Remember, it's not magic, it's science!

Materials you'll need:

- Pencils (the sharper the better!)
- Ziploc bags
- Water
- Paper towels
- An area outside or a sink (to practice)

What to do:

1. Fill your Ziploc bag 2/3 full with tap water.
2. Take it outside or continue the experiment over the sink.
3. Hold the water-filled bag by the top in one hand, and a sharpened pencil in one hand.
4. Slowly, but firmly, push the pencil in and through the bag (and keep it there!)
5. Now push another pencil through the other side of the bag.
6. Keep going with the rest of the pencils!
7. When you're done, hold the bag over the sink or lawn and remove the pencils. The water will come gushing out.

What's happening?

Your Ziploc plastic bags is made of *polymers*, which are long chains of individual molecules, called *monomers*. When you puncture the bag with the pencil, you're separating the bag's *polymer chains* without breaking them. The long chains of molecules than squeeze in tight around the surface of the pencil preventing any sort of leak.

63) PENNY DROPS

How many drops of water can fit on a penny? Use this experiment to find out!

Materials you'll need:

- A Penny
- Paper towel
- Flat surface
- Eyedropper
- Water

What to do:

1. Wash and rinse a penny in tap water.
2. Dry it completely with a paper towel.
3. Place the penny on a flat surface.
4. Use an eyedropper to collect some water.
5. Carefully drop individual drops of water onto the flat surface of the penny.
6. Count and record the number of water drops as you add them one at a time, until water runs over the edge.

What's happening?

This experiment involves *cohesion* and *surface tension,* which is the act of water molecules sticking to each other. Water's *cohesion* and *surface tension* becomes obvious when the drops of water reach the edge of the penny. Here you can see a dome of water forming on top of the penny, which is the water molecules clinging to each other, before it becomes too much and they spill over the edge.

64)PRESSURE ICE CUBES

Cut a piece of ice in half like magic while learning about pressure and how it relates to ice skating.

Materials you'll need:

- An ice cube
- A piece of fishing line with a weight tied to each end
- A container
- A plate or tray to keep things clean and dry

What to do:

1. Turn the container upside down and place it on the tray.
2. Place the ice cube on top of the upside down container.
3. Rest the weighted fishing line over the ice cube so that the ends with the weights are left dangling on either side of the container.
4. Wait and watch for the next 5 minutes or so.

What's happening?

The pressure from the two weights is responsible for this experiment. The weights pull the string through the ice cube, which melts the ice directly under the fishing line. This process is similar to what happens in ice skating—the blades of a skater are like the fishing line…they melt the ice directly underneath, allowing them to move smoothly on a thin layer of water.

65)MYSTERIOUSLY MOVING WATER

Water can move in mysterious ways, learn why and how with this fun science experiment.

Materials you'll need:

- Two drinking glasses
- Water
- Some paper towels

What to do:

1. Fill one glass with water, keep the other one empty. Place them next to each other.
2. Twist a couple of pieces of paper towel together. It should look a little like a piece of rope.
3. Place one end of the paper towels into the glass filled with water and the other into the empty glass.
4. Be patient and watch what happens.

What's happening?

Your paper towel rope should start getting wet, and after a few minutes you will see that the empty glass is starting to fill with water. It will keep filling up until there is an even amount of water in each glass.

This happens because of a process called *capillary action*, which is the same process that occurs in plants when moisture travels from the roots to the rest of the plant. In this experiment, water move along the tiny gaps in the fibers of the paper towels. This happens because the *adhesive force* between the water and paper towel is stronger than the *cohesive forces* inside the water alone.

66)WEIGHTLESS H$_2$0

This experiment isn't magic, but it will seem like it when you make a glass of weightless water (H_2O). You may want to try this experiment outside or over the sink until you get really good.

Materials you'll need:
- A drinking glass
- Water
- A piece of cardboard

What to do:

1. Fill the glass of water all the way to the top.
2. Place the cardboard over the top of the glass (be sure that air bubbles enter the glass).
3. Hold the cardboard on tight and turn the glass upside down.
4. Take away your hand holding the cardboard.

What's happening?

The cardboard and water should stay put like magic! Even though the glass is upside down, the water stays in place. How can this be? When you put the cardboard over the full glass of water, it stops air from entering the glass. This makes the air pressure around the glass greater than the pressure of the water inside the glass. It's this air pressure that keeps the cardboard in place and the water inside the glass.

67) BURPING BOTTLE

This experiment will show you that burps really are just swallowed air!

Materials you'll need:

- Empty glass bottle
- Dime
- Glass of water
- Straw
- Freezer

What to do:

1. Place a dime over the mouth of the bottle.
2. Place the bottle (with the dime covering the mouth) inside a freezer.
3. Wait several hours for the bottle to get as cold as possible, then take it out.
4. Place one end of a straw into a glass of water and cover the other end with your thumb.
5. Withdraw the straw from the water (with your thumb in place). Water will be trapped inside the straw.
6. Position the water-filled straw over the dime-covered bottle mouth.
7. Remove your thumb and let water to pour onto the dime.
8. Watch and listen!

What happens?

The dime should be lifted off the mouth of the bottle by a burp from inside! This happens because of water and air. When you place the dime over the mouth of the bottle, you create a seal, and the air in the bottle stays in the bottle. By putting the bottle in the freezer, the air particles inside of the bottle slow down and become *denser*. This means that there is *low air pressure* inside the bottle. When you take the bottle out and the air inside it begins to warm, it allows the air inside the bottle to expand which raises the air pressure inside. When you drop water onto the dime, it creates another seal through *cohesion* and *adhesion*. *Cohesion* is the act of water molecules clinging to each other, and *adhesion* is the act of the water molecules clinging to the mouth of the bottle. The increased air pressure pushes against the watery seal, lifts the dime up off the bottle, and escapes with a loud burp!

68) TOOTHPICK SUBMARINE

Materials you'll need:

- Wooden toothpick (with one flat side and one sharp side. If you don't have those types of toothpicks, simply snip off one sharp end).
- Pan of water
- Shampoo

What to do:

1. Dab a little shampoo onto the flat end of your toothpick.
2. Drop the toothpick into your pan of water and watch!

What's happening?

The toothpick should start moving in the direction of its sharp end. This is because shampoo contains ingredients that reduce the *surface tension* of water. As the shampoo on the toothpick dissolves, it reduces the water's *surface tension*, and releases the water's hold on that end of the toothpick. The surface tension of the water is still strong on the other side of the toothpick, so it pulls the toothpick in that direction.

69) HARD & SOFT WATER

"Hard water" means that it contains minerals. These minerals can interfere with the cleaning ability of soaps and detergents, which is why many people have "water softeners." Water softeners remove these minerals. In this science experiment, you will compare how soap and suds work in soft and hard water.

Materials you'll need:

- 2 cups of distilled water
- 1 teaspoon Epsom salts
- 2 clean and empty 2-liter plastic bottles, with screw caps
- Permanent marker
- several drops of liquid dishwashing detergent (NOT the kind for dishwashing machines)

What to do:

1. Pour 1 cup of distilled water into each of the empty plastic bottles.
2. Add the 1 teaspoon of Epsom salts to one of the bottles. Swirl the bottle until the Epsom salts dissolve.
3. Label the bottle with Epsom salt in it, "hard water." Label the other bottle "soft water"
4. Add several drops of liquid dish detergent to both bottles. Seal the bottles with their caps.
5. Shake both bottles.
6. How are they different?

What's happening?

A large amount of suds will form in the "soft water" bottle.
Far fewer suds will form in the "hard water" bottle.
In this experiment, you made "hard water" by adding
Epsom salt, which is *magnesium sulfate*. Magnesium in
water interferes with the cleaning action of soaps. Instead it
combines with soap and forms a "soap scum" that does not
dissolve in water.

70)ICE MELTING EXPERIMENT

Do you think that an ice cube sitting at the top of a glass will melt and spill over the sides? Complete this experiment to find out! This experiment uses warm water, so ask for help from an adult for this.

Materials you'll need:

- A clear glass
- Warm water (ask for help!)
- An ice cube

What to do:

1. Fill your glass all to the top with warm water.
2. Gently lower the ice cube into the water. (Try not to touch the glass when you do this).
3. As the ice cube melts, watch the water level carefully. What happens?

What's happening?

Even though the ice cube melts, the water shouldn't overflow. When water freezes, and become ice, it expands and takes up more space than it does when it is a liquid. When the ice cube melts, it takes up less space than it did as a piece of ice and the water level stays about the same.

71)ICE FISHING

Have you ever wondered why people sprinkle salt on the road when it gets icy? Use this experiment to find out!

Materials you'll need:

- Cup of water
- A few ice cubes
- Table salt
- A piece of string

What to do:

1. Place the ice cubes in the cup of water.
2. Use your string to try and "fish" for an ice cube. (You'll notice that you don't catch anything).
3. Now place the string across the top of the ice cubes.
4. Sprinkle some salt over the ice cubes.
5. Wait 1 minute.
6. Now pull on the string and see what you've caught!

What's happening?

When you mix salt with ice, you lower the *freezing point*. The normal freezing point for water is 32° F (0° C), but when you add salt, it lowers the freezing point and helps the ice melts more easily, which allows your string to "catch it."

72)ICE POWER!

In this experiment you'll learn about the expanding power of ice.

Materials you'll need:

- three small plastic drinking cups (one with a lid)
- a large dish
- water
- freezer

What to do:

1. Fill all three cups with as much water as possible without overflowing.
2. Put the lid on one of the cups.
3. Set all three cups on the dish and place in the freezer overnight.
4. Check to see what happened the next morning.
5. Leave the cups in the freezer for a few days.
6. Did you notice any changes?

What's happening?

When water freezes and becomes ice, it *expands* (which means it takes up more space as a solid than it does as a liquid). This is why when water is left in a garden hose when it is freezing outside, the force of the ice expanding can cause the hose to break open or the pipes to burst.

73)MAKING MUSIC WITH WATER

Have you ever tried making music with water? In this experiment, you can make some cool music by turning glasses of water into instruments.

Materials you'll need:

- 5 (or more) drinking glasses (or glass bottles)
- Water
- Wooden stick (a pencil also works)

What to do:

1. Place the glasses next to each other and fill them with different amounts of water. Put just a little bit of water in the first glass, and then a little bit more in each one, until the last glass is almost full.
2. Use the wooden stick and hit the side of glass with the least amount of water and listen to the sound.
3. Now hit the glass with the most water and listen to the sound.
4. Hit the other glasses and see what noise they make. Try to see if you can make a song by hitting the glasses in a different orders.

What's happening?

Each of the glasses will have a different *tone* when hit with wooden stick. The first glass (with the least amount of water) will have the highest tone, while the glass with the most water will have the lowest. This happens because when you hit the glass, you make small vibrations which creates sound waves. These sound waves travel through the water and more water means the vibrations go more slowly and create and deeper (lower) tone.

74)MAKE A PAPER CLIP FLOAT

Materials you'll need:

- paper clips
- tissue paper
- a bowl of water
- pencil with eraser

What to do:

1. Fill the bowl with water
2. Place a paper clip in the bowl of water. What happens?
3. Now, tear a piece of tissue paper (about half the size of a dollar bill).
4. Drop the tissue flat onto the surface of the water.
5. As gently as possible, place a dry paper clip flat onto the tissue (try not to touch the water or the tissue)
6. Use the eraser end of the pencil to carefully poke the tissue (not the paper clip) until the tissue sinks.
7. What happens?

What's happening?

This is happening because of *surface tension.* The paperclip is not actually floating, it is being held up by the surface tension, which means the water molecules are holding on tight to each other.

75)WATER EXAMINATION EXPERIMENT

Water not only keeps us alive, it also can store life! Pond water, water near plants, and plain old tap water can be home to a lot of cool creatures (microorganisms). In this experiment, we'll take some samples and view them under a microscope.

Materials you'll need:

- A concave slide
- A dropper
- A microscope
- Different samples of water (tap water, muddy water, pond water…the dirtier the water, the more microorganisms there will be!)

What to do:

1. Set up the microscope and use its highest setting.
2. Using the dropper, take some water from one of your samples and put it on the slide.
3. Focus the microscope. What do you see? (Be patient and check the focus of the microscope if you can't see anything. If you still can't see anything, try a different sample of water).

What's happening?

The various microorganisms you may see in the water include:

- **Euglenas:** These have a long tail called a *flagella* which allows them to move.
- **Protozoa:** These also have a flagella, though it is quite small. They appear very similar to algae.
- **Amoebas:** These microorganisms swim by wobbling. When they eat something, they surround it like a blob.
- **Algae:** These organisms range from looking yellow to green to red. You may see them by themselves or in chains.

What else can you see under your microscope?

VIII. STATIC ELECTRICITY

76)CREATE STATIC ELECTRICITY (METHOD 1)

Static electricity is fascinating, and this experiment will teach you about positively and negatively charged particles, while also showing you why opposites attract.

Materials you'll need:

- 2 inflated balloons
- Wool fabric
- A head of hair (your own works great!)

What to do:

1. Rub the 2 balloons one by one against the woolen fabric, then try moving the balloons together. Do they rub together easily or do they resist each other?
2. Rub 1 of the balloons back and forth on your hair really quickly, then slowly it pull it away. Looking in a mirror to see what happens! (Or if there is no mirror, ask someone nearby to tell you what is happening)

What's happening?

When you rub the balloon against your hair or the woolen fabric, invisible *electrons* (particles with a negative charge) build up on the surface of the balloon. This is called *static electricity,* which means non-moving electricity.

In the first experiment, both the balloons were negatively charged after rubbing them against the woolen fabric. Because they were the same, they were unattracted to each other. This is why they resisted coming together.

Part 2 of this experiment shows you why opposites attract. The *protons* (particles with a positive charge) on the surface of your hair are attracted to the negatively charged balloon. This is why your hair starts to rise up to meet the balloon when you pull it away.

77)CREATE STATIC ELECTRICITY (METHOD 2)

This experiment shows you how to move things with the power of static electricity (and teaches you a little bit more about how opposites attract).

Materials you'll need:

- An empty aluminum can (soda cans works great!)
- 1 inflated balloon
- A head of hair

What to do:

1. Place the empty soda can on its side on a flat smooth surface (like a table or floor without carpet).
2. Rub the balloon back and forth on your hair a few times.
3. Now, hold the balloon close to the can without it actually touching the can. What happens?

What's happening?

The can should start rolling towards the balloon without you even touching it! Like the first static electricity experiment, when you rub the balloon on your hair, invisible *electrons* (negatively charged particles) build up on the surface of the balloon. When you place the balloon near the can, those electrons have the power to pull the very light can (with a positive charge) towards them. Once again showing that opposites attract.

MORE: Make some small changes and see what happens! If you change the size of the balloon does it affect the power of its pull? Does the length of your hair affect the power of the static electricity? Try putting water in the can—how much can you put it until it is too heavy for the balloon to pull?

78)WATER BENDING EXPERIMENT

Learn how to "bend" water with static electricity!

Materials you'll need:
- A plastic comb
- A sink faucet
- A head of hair

What to do:

1. Turn on the water so it is coming out in a narrow stream.
2. Run the comb through your hair 10-15 times.
3. Take the comb and slowly move it towards the stream of water (without touching it), and see what happens!

What's happening?

Like our other static experiments, the static electricity you built up by combing your hair attracts the stream of water, making it "bend" towards the comb. When you rub the comb through your hair, *electrons* (negatively charged particles) move from your hair to the comb. This means the comb now has extra electrons and is "negatively charged". The water has both positively and negatively charged particle, and its positive charges are attracted to the negative charges on the comb (remember, with static electricity, opposites attract!). The comb's electrons attracts the water's positively charged particles, and is actually strong enough to bend the water towards the comb.

79)MAKE A STATIC GHOST

Materials you'll need:

- A piece of tissue paper
- A balloon
- Scissors
- A head of hair

What to do:

1. First cut out a ghost shape in the tissue about 1 ½ inches long and add some eyes with a marker.
2. Blow up the balloon and tie it.
3. Rub the balloon really fast through your hair for about 10 seconds.
4. Slowly bring the balloon near the ghost, and the ghost will begin to rise toward the balloon. With a little practice, you can get the ghost to rise, float, and even dance around.

What's happening?

When you rub the balloon through your hair, invisible *electrons* (with a negative charge) build up on the surface of the balloon. The electrons have the power to pull very light objects (with a positive charge) toward them.

*Tip: The best way to make the ghost rise without it sticking to the balloon is to tape the very tip of the bottom of the ghost to a table.

80)CREATE A LEVITATING ORB

Materials you'll need:
- 1" wide PVC Pipe about 2 feet long.
- Thin Mylar tinsel
- A head hair
- Scissors

What to do:

1. Arrange 6 strands of Mylar together and tie them together in a knot at one end.
2. Tie them together again about 6 inches from the first knot.
3. Trim off the loose ends of the Mylar
4. Charge the PVC pipe by quickly rubbing it back and forth through your hair for 10 seconds.
5. Hold the Mylar orb by one of its knots above the charged pipe and let it drop.
6. What happens?

What's happening?

It is all about *static electricity*. Similar static charges repel away from each other. When you rub the pipe in your hair you give the pipe a negative static charge. The orb is attracted to the pipe at first because the orb has a positive charge. As soon as the orb touches the pipe, it picks up a negative charge. Since the pipe is negative and the tinsel orb is now negative, they repel away from each other and the orb levitates!

81) ELECTRIC CORNSTARCH

In this experiment, you will use static electricity to make cornstarch "electric!"

Materials you'll need:

- Cornstarch
- Vegetable oil
- Mixing bowl
- Large spoon
- Balloon
- Head of hair or carpet
- Measuring cup

What to do:

1. Pour ¼ cup cornstarch into your mixing bowl.
2. Add ¼ cup vegetable oil to the cornstarch. Mix well until it becomes thick.
3. Now blow up a balloon tie it.
4. Rub the balloon on your head quickly for 10 seconds (or on the carpet).
5. Take a spoonful of the cornstarch mixture and hold it close to the balloon.
6. Once the cornstarch jumps toward the balloon, slowly move the balloon away.
7. See how far the cornstarch will jump!

What's happening?

When you rub the balloon your hair, you give the balloon *electrons* (negatively charged particles), which generate a *negative static charge*. However, the cornstarch has a *neutral charge* (it has both positive and negative particles). The negative charge of the balloon will repel the electrons of other objects while attracting its protons. The negatively charged object will attract a lightweight object, which is why you drip cornstarch from a spoon—it makes it lighter! This allows the dripping cornstarch to swing towards the balloon more freely.

82) FLOATING STATIC BAND

This experiment teaches electrostatic propulsion and the repulsion of like charge.

Materials you'll need:

- A wool sweater
- Thin sheet of Styrofoam (packing wrap)
- A piece of PVC pipe (3/4-inch diameter, 3-feet long)
- A pencil
- Tape
- Glue (or stapler)

What to do:

1. Tape a pencil onto the end of the PVC pipe.
2. Cut the thin Styrofoam sheet into strips that are 1"wide x 12" long.
3. Make a ring out of the Styrofoam strips by joining the two ends and securing them with glue or staples.
4. Rub the Styrofoam band against wool sweater to build up a static charge.
5. Place the "charged up" band on the table.
6. Now rub the wool up and down on the PVC pipe to build up its static charge.
7. Use the pencil on the end of the PVC pipe to pick up the band and toss it in the air. Quickly position the pipe underneath the band to make it float. (This step may some practice, but don't give up—it's worth it!)

What's happening?

Rubbing the wool against the Styrofoam band and the PVC
pipe transfers *electrons* (negatively charged particles) to
both objects. The band floats above the pipe because the
similar charges repel one another. This is called
electrostatic propulsion and the repulsion of like charge.
Whew!

83) STATIC ZINGERS

Materials you'll need:

- Clean, plastic 1-liter soda bottle with cap
- ¼ cup Styrofoam beads
- A wool sweater (or a head of hair)

What to do:

1. Take your clean plastic bottle and fill it with the Styrofoam beads and seal the bottle with a cap.
2. Rub the bottle on your head or on a wool sweater.
3. Watch the effects of static electricity on the beads.
4. Now run your hand over the plastic bottle.
5. What do the static beads do?

What's happening?

When you rub the plastic bottle through your hair, invisible *electrons* (with a negative charge) build up on the bottle and pass through into the Styrofoam beads, causing them to move.

IX.   EXPERIMENTS WITH YEAST (IT'S ALIVE!)

84) YEAST BALLOON

This is similar to the **Balloon Blow-Up Experiment**, but this time we will be using different materials—so read on and have fun learning about what yeast can do!

Materials you'll need:

- A packet of yeast
- A small, clean, clear, plastic soda bottle (16 oz. or smaller)
- 1 teaspoon of sugar
- Some warm water
- A small balloon

What to do:

1. Fill the bottle up with about one inch of warm water.
2. Add all of the yeast packet and gently swirl the bottle a few seconds.
3. Add the sugar and swirl it around some more.
4. You may not be able to see it, but the yeast is eating the sugar!
5. Blow up the balloon a few times to stretch it out then place the neck of the balloon over the neck of the bottle.
6. Let the bottle sit in a warm place for about 20 minutes

What's happening?

As the yeast eats the sugar, it releases a gas called carbon dioxide (CO_2). The gas fills the bottle and then fills the balloon as more gas is created.

85) USE CHEMISTRY TO MAKE HEAT

This is a great chemical reaction experiment to learn about heat!

Materials you'll need:

- 1 tsp yeast
- 1/2 cup hydrogen peroxide
- stirring stick
- thermometer

What to do:

1. Record the temperature of the hydrogen peroxide and place it in a small bowl.
2. Add the dry yeast to the peroxide and stir
3. Watch for changes in the mixture and the temperature

What's happening?

Hydrogen peroxide (H_2O_2) is mostly water with an extra oxygen molecule. The yeast releases the oxygen, creating the bubbles, and it also releases heat (an *exothermic reaction*.)

86) ELEPHANT TOOTHPASTE

This project is a fun experiment exploring different reactions! It requires warm water, so be sure to ask an adult for help.

Materials you'll need:

- 2 Tablespoons of warm water (ask for help!)
- 1 teaspoon yeast
- Large pan
- Empty plastic bottle
- ½ cup of 6% hydrogen peroxide (anything under 6% won't work).
- 4-5 drops food coloring
- 1 squirt of dish soap

What to do:

1. Set the empty plastic bottle in the middle of a pan.
2. Combine the warm water and yeast in a small bowl.
3. Mix together for 1 minutes.
4. Add the hydrogen peroxide, food coloring, and dish soap in the plastic bottle.
5. Pour the yeast mixture into the plastic bottle and watch!

What's happening?

Yeast breaks down hydrogen peroxide into water and oxygen. The squirt of dish soap you add catches the oxygen to make larger bubbles, and you get elephant toothpaste! If you touch the foam or bottle it will feel warm—this is because the chemical reaction is *exothermic* (it releases energy as heat).

X.   HEAT, LIGHT, AND ENERGY

87) HOT SUGAR EXPERIMENT

Learn more about temperature and water molecules with this sugar experiment. This experiment involves hot water, so be sure to ask for an adult's help.

Materials you'll need:

- Sugar cubes
- Cold water in a clear glass
- Hot water in a clear glass (ask for help with this!)
- Spoon

What to do:

1. Be sure that the glasses have an equal amount of water.
2. Put 1 sugar cube into the glass of cold water and stir with the spoon until the sugar dissolves.
3. Add another sugar cube and repeat the process.
4. Keep adding sugar cubes (and remember to count how many you add) one at a time, and stirring, until the sugar stops dissolving (when the sugar starts to gather on the bottom of the glass).
5. Record how many sugar cubes you could dissolve in the cold water.
6. Now repeat the same process for the hot water.
7. At the end, compare the number of sugar cubes dissolved in each cup.

8.   Which one dissolved more?

What's happening?

The hot water dissolved more sugar than the cold water.
Why? The hot water dissolves more sugar, because it the
heat causes it to have faster moving molecules. These
"fast" molecules are spread further apart than the molecules
in cold water. Because there is more space in between the
hot-water molecules, more sugar molecules can fit in
between, and dissolve.

88)HEAT ABSORPTION EXPERIMENT

When you're out in the sun on a hot day you get hotter faster if you wear dark colors. Why is that? Complete this experiment to find out.

Materials you'll need:

- 2 clear glasses or jars (the same size)
- Water
- Thermometer
- 2 rubber bands or scotch tape
- 1 piece of white paper
- 1 piece of black paper

What to do:

1. Wrap the white paper around one of your glasses. Secure it with the rubber band or the tape.
2. Repeat the process with the black paper and the other glass.
3. Fill the glasses with the same amount of water.
4. Leave the glasses out in the sun for 2-3 hours.
5. Return and take the temperature of each glass of water.
6. Record your results. How are they different?

What's happening?

Feel the glass with the black paper around it—it should feel hotter than the other. This is because the black paper absorbs more light and heat than the white paper. Lighter surfaces *reflect* more light, whereas dark surfaces *absorb* it. That's why people try to wear lighter colored clothes when it is hot outside.

89) HOT AIR EXPERIMENT

Air acts differently, depending on temperature. Use this experiment to find out what happens to a balloon when the air inside it heats up!

Materials you'll need:

- Empty bottle
- Balloon
- Pot of warm (NOT boiling) water (Ask for help from an adult with this).

What to do:

1. Stretch the balloon a few times with your hand to make it flexible.
2. Now stretch it over the mouth of the empty bottle.
3. Put the bottle in the pot of hot water.
4. Wait a few minutes and watch what happens!

What's happening?

As the air inside the bottle and balloon heats up, it starts to expand. The molecules begin to move faster and further apart from each other. This is what makes the balloon stretch. There is still the same amount of air inside both the balloon and bottle, it has just *expanded* (gotten bigger) as it heats up. Now you know that warm air takes up more space than the same amount of cold air.

90)MAGNIFIED BALLOON

This experiment demonstrates the power of the sun, magnified! Magnifying the sun can be dangerous, so ask an adult for help.

Materials you'll need:

- Clear balloon
- Black balloon
- Magnifying glass
- Sunlight

What to do:

1. Blow up the clear balloon, but just hold the end shut, do not tie it up yet.
2. Partially insert the black balloon into the clear balloon, with the mouth of the black balloon sticking out of the clear balloon.
3. Now blow up the black balloon until it is about half the size of the other balloon.
4. Tie off the black balloon, and push it the rest of the way into the clear balloon.
5. Now tie off the clear balloon.
6. With the help of an adult, use the magnifying glass to focus sunlight on the black balloon.
7. Watch what happens!

What's happening?

The black balloon inside should burst! When you use a magnifying glass you focus the sun's rays into a dot that is incredibly hot. This magnified dot can get so hot that it can start a fire! Only the black balloon bursts because most of the light and its heat can pass right through the clear balloon's surface. The black balloon isn't so lucky, because it doesn't reflect any light. Instead, it absorbs almost all of the light and its heat which causes it to burst.

91)HOW FAST DO MOLECULES MOVE?

This is another experiment to test the speed of water molecules, with some great visual results! This experiment involves hot water, so be sure to ask for an adult's help.

Materials you'll need:

- Cold water in a clear glass
- Hot water in a clear glass (ask for help with this!)
- Food coloring
- An eye dropper

What to do:

1. Make sure that the glasses are filled with the same amount of water (one cold and one hot).
2. Quickly put one drop of food coloring into each of the glasses.
3. Pay attention and watch what happens!

What's happening?

If you watch carefully, you'll notice that the food coloring spreads faster throughout the hot water than in the cold. This is because the water molecules in the hot water move at a faster rate, which helps to spread out the food coloring more quickly than the slower-moving cold water molecules.

92) ENERGY TRANSFER

Energy is always moving; do this experiment to see for yourself!

Materials you'll need:

- A basketball or soccer ball
- A tennis ball
- A large space outside

What to do:

1. Carefully place the tennis ball on top of the bigger ball, and hold it in place.
2. Let go of both the balls at exactly the same time.
3. Watch what happens!

What's happening?

The tennis ball should bounce off the bigger ball and fly up into the air. When the bigger ball hits the ground, all of the *kinetic energy* in the bigger ball is transferred through to the smaller tennis ball, which sends it shooting high into the air. When you were holding both of the balls in the air, they had another type of energy called *potential energy,* which is the energy you used to hold them up. This experiment shows you that energy is never lost, it is only transferred from one thing to another.

93)MARSHMALLOW SLINGSHOT!

This fun experiment is all about the *transformation of energy*…..and, of course, FUN!

Materials you'll need:

- Rubber bands
- The plastic ring from the lid of milk jug (or from a prescription bottle)
- Bag of marshmallows
- A chair

What to do:

1. Make a chain of rubber bands with the ring in the center. You can do this by overlapping two rubber bands, pull the bottom one through the one on top, and then through itself. Just keep going until the chain is long enough to your liking.
2. Turn your chair upside down, and secure your rubber band catapult its legs.
3. Set up a few targets around, and place and marshmallow in your catapult, pull back, and let it fly!

What's happening?

This crazy experiment demonstrates *Newton's Third Law*: Every action has an equal and opposite reaction. When you use force to pull back on the marshmallow, it flies forward once you release it. This is also a great activity for *transformation of energy.* When you pull the rubber band back, you are supplying it, and the marshmallow, with *potential energy* (stored energy). Then, when you release it, the *potential energy* turns into *kinetic energy* (the energy of motion). No energy is lost, it is just *transformed* from one kind to another.

94) FLOATING SOAP BUBBLES

Soap bubbles can do many different things under different conditions. In this test, examine them closely and see what your soap bubbles reveal.

Materials you'll need:

- soap bubble solution
- a wand for blowing soap bubbles
- a large transparent container with an open top
- ½ cup of baking soda
- 1 cup of Vinegar
- shallow glass dish to fit inside large container

What to do:

1. Place your glass dish on the bottom of the large transparent container.
2. Put the baking soda in the glass dish.
3. Add the vinegar (it should immediately start to fizz)
4. After 1 minutes (once the fizzing in the dish has stopped), gently blow several soap bubbles over the opening of the large container, so that they settle into the container. What happens?
5. While the bubble is floating in the container, you can observe the soap bubble closely. Note what the bubble looks like—what color is it? Can you see more than one color on the bubble? Do the colors change? Notice the size of the bubble. Does its size change? Does it rise or sink in the container?

What's happening?

The colors of a soap bubble come from reflections of the white light that falls on the bubble. White light, such as from the sun or from a light bulb, contains light of all colors. Light has waves, and the length of the wave, from crest to crest, determines the color of the light. When light reflects from a bubble, some of each wave reflects at the outside surface of the soap film. Some light travels through the soap film, and reflects from the inside surface of the film.

Interference between waves occurs whenever waves travel through the same space. Interference occurs when two rocks are tossed near each other into a lake. Circular waves on the surface of the water spread out from where each rock entered the water. Where the crests of two waves meet, interference between the waves causes the motion of the surface of the water to increase. Where a crest and a valley meet, interference reduces the motion of the water's surface. Similar interference can occur in waves of light.

95)IVORY FOAM

This soap experiment demonstrates closed-cell foam formation and physical change. Have fun!

Materials you'll need:

- 1 bar of Ivory soap
- paper towel
- microwave

What to do:

1. Place the unwrapped bar of Ivory soap on a paper towel.
2. Nuke your soap. Watch the soap closely to see what happens.
3. Depending on microwave power, your soap will reach its maximum volume within 90 seconds- 2 minutes.
4. Allow the soap to cool for a minute or two before touching it.
5. The soap will feel brittle and flaky, but it's still soap, with the same cleaning power as before. Go ahead and get it wet and you'll see it lathers the same as ever.

What's happening?

In the microwave, you are not only heating the soap, you are also heating the air and water trapped inside it, which causes the water to *vaporize* and the air to *expand*. The *expanding gases* push on the softened soap, which makes it also expand and become a foam—the appearance of the soap changes, but no chemical reaction occurs. This is called a *physical change.* This experiment also is a great demonstration of *Charles' Law*, which states: the volume of a gas increases with its temperature. When the water and air molecules inside are heated, their volume expands and the soap becomes a puffy soufflé. Ivory is the best soap for this experiment, because it contains a lot of whipped air inside.

96) THERMITE REACTION

This simple experiment only uses aluminum foil and old ball bearings to create a brilliant chemical reaction. Get adult help with this experiment.

Materials you'll need:

- Rusted iron or steel ball bearings
- Aluminum foil

What to do:

1. Wrap a rusted steel ball bearing completely with aluminum foil.
2. Now strike the foil-covered ball with a second rusted, but uncovered, ball bearing.
3. What happens? (After striking one spot, you'll need to rotate the foil-covered ball to strike it in another spot).

What's happening?

A *thermite reaction* is an *exothermic oxidation-reduction reaction*. To make a thermite reaction you need heat, fuel, and *metal oxide*. In this experiment, the fuel is aluminum foil, and the metal oxide is the rust on the ball bearings (aka *iron oxide*). When you hit them together you create heat, and you have all of the elements needed to create a thermite reaction!

97) RUBBER BAND HEAT

In this experiment you will examine the *thermal properties* of rubber. Which means, you going to see how rubber bands behave with heat (a form of energy).

Materials you'll need:

- 1 heavy-duty rubber bands
- Yourself!

What to do:

1. Place your thumbs through a rubber band, one on each end. Without stretching the band, hold it to your forehead or lip. Does the band feel cool or warm or about the same as your skin?
2. Now, move the rubber band away from your face, so it is not touching your skin. Quickly stretch it as far as you can and (holding it in the stretched position) touch it again to your face again. Does it feel warmer or cooler or about the same?
3. Move the stretched rubber band away from you face. Quickly let it relax to its original size and again hold it to your skin. Does it feel warm or cool?
4. Repeat experiment, relaxing, stretching, and testing the rubber band several times until you are sure of the results.

What's happening?

An object feels cool to your skin when heat flows from your skin to the object. On the other hand, an object feels warm when heat flows from the object to your skin. If the stretched rubber band feels cool, it is because it is absorbing heat from your skin. If it feels warm, then it is giving off heat to your skin.

98)MAKE A FIREPROOF BALLOON

A fire can weaken the rubber in a balloon and cause it to pop. But with this science experiment, you will find out how you can hold a balloon directly in a flame without it popping. This experiment requires the use of fire, so be sure to ask an adult for help.

Materials you'll need:

- 2 round balloons, not inflated
- Several matches (ask for help!)
- ¼ cup water

What to do:

1. Inflate one of the balloons and tie it closed.
2. Pour the water in the other balloon, and then inflate it, and tie it shut.
3. Light a match and hold it under the first balloon. What happens?
4. Light another match. Hold it directly under the water in the second balloon. Allow the flame to touch the balloon. What happens?

What's happening?

The flame heats whatever it is around, so it heats the rubber of both balloons. The rubber of the first balloon (without water) becomes so hot, that it becomes weak and cannot resist the air pressure inside. The second balloon has water in it, and when it is near the flame, the water absorbs most of the heat. This makes it so the rubber of the balloon does not get so hot, so it doesn't get weaker, which causes it not to break.

99)MAKE AN ELECTROMAGNET

Materials you'll need:

- A large iron nail
- About 3 feet of thinly coated copper wire
- One D battery
- Some paper clips or other small magnetic objects

What to do:

1. Leave about 8 inches of wire loose at one end and wrap most of the rest of the wire around the nail. Try not to overlap the wires.
2. Cut the wire so that there is about another 8 inches loose at the other end too.
3. Now remove about an inch of the plastic coating from both ends of the wire and attach the one wire to one end of a battery and the other wire to the other end of the battery.
4. Now you have an electromagnet!
5. Put the point of the nail near a few paper clips and it should pick them up!

What's happening?

Most magnets, like the ones on many refrigerators, cannot be turned off, they are called permanent magnets. Magnets like the one you made that can be turned on and off, are called electromagnets. They run on electricity and are only magnetic when the electricity is flowing. The electricity flowing through the wire arranges the molecules in the nail so that they are attracted to certain metals.

100) "MAGIC" STRAW

In this experiment, bend a straw without even touching it!
Though it sounds like magic, it's really just an amazing
scientific principle at work.

Materials you'll need:

- A glass
- Water
- A drinking straw

What to do:

1. Fill your glass with water until it is half full.
2. Put the straw in the glass.
3. Look at the straw from the top and bottom of the glass.
4. Now, look at the straw from the side of the glass,
   focusing on the point where the straw enters the water.
5. What do you see?

What's happening?

Our eyes are use light to see, but when light travels through
different objects (like water and air), it changes direction.
Light *refracts* (bends) when it passes from water to air.
This makes the straw appear like it is bending, because you
are seeing the bottom half through water and air, but the top
half through air only.

101) BEND LIGHT

Materials you'll need:

- Clear plastic bottle
- Thumb tack
- Nail
- Water
- Flashlight
- Dark room
- Sink

What to do:

1. Using the thumb tack, punch a hole into the plastic bottle, 2 inches from the bottom. Make the hole larger with the nail.
2. Place your finger over the hole, and fill the bottle with water. Put the cap on and place the bottle on the edge of the sink (with the hole facing the sink).
3. Turn the lights off and with your other hand (or with a helper) turn the flashlight on and hold it.
4. Take the cap off the bottle and aim your flashlight (through the plastic bottle) at the stream of water. What do you see?

What's happening?

The light should bend with the arc of water and create a bright glow where both the water and the light hits the sink.

When the light in the stream strikes the border between the stream of water and air, most of the light is reflected back into the stream. The light continues this *internal reflection* all the way down into the sink. This is the same principle that applied to transmit light signals through optical fibers.

102) X-RAY VISION

Materials you'll need:

- Black marker (like a Sharpie)
- Brown envelope
- White Envelope
- Dark piece of construction paper

What to do:

1. Using the black marker, write a three-letter word in large letters on the white piece of paper.
2. Place the paper in the brown envelope (without folding it).
3. Now place the brown envelope in the white envelope. It should be impossible to read what you wrote.
4. Roll your piece of dark construction paper into a tube (about 4 inches long).
5. Place it up against the envelope and look through it. What do you see?

What's happening?

It is difficult to see the writing inside the white envelope because of the light reflected off its surface. But when you use the dark tube to look through, it blocks the reflected light, so you only see the light coming through the envelope and can read what you wrote.

103)   MAKE A RAINBOW!

Do you like rainbows? Now you can find out how they work by making your own, using just a few simple items from around the house.

Materials you'll need:

- A drinking glass
- Water
- White paper
- A sunny day

What to do:

1. Fill your drinking glass until it is ¾ of the way full.
2. Take the glass of water and the white paper an area of your house with plenty of sunlight (near a window is best).
3. Carefully hold the glass of water above the paper and watch as sunlight passes through it. What does it form on the white paper?

What's happening?

Rainbows form in the sky when sunlight passes through raindrops and *refracts* (bends). The exact same thing happens when it passes through your glass of water. The sunlight bends and separates into the colors of the rainbow—red, orange, yellow, green, blue, indigo and violet.

MORE: Move the glass around to get different rainbow effects. Hold it at different heights and angles and record your results.

104)   SUNSET MILK

This experiment will show you why the sky looks blue during the day, but red when the sun is rising or setting.

Materials you'll need:

- a flashlight
- a large, clear container (a 2½-gallon aquarium is perfect)
- 1 cup of milk

What to do:

1. Set the container on a table where you can view it from all sides.
2. Fill it ¾ full with water.
3. Turn on the flashlight and hold it against the side of the container so its beam shines through the water. Try to see the beam as it shines through the water (it may be rather difficult to see exactly where the beam passes through the water).
4. Add ¼ cup of milk to the water and stir it.
5. Like before, hold the flashlight to the side of the container. Notice that the beam of light is now easily visible as it passes through the water. Look at the beam both from all angles. (From the side, the beam appears bluish, and on the end, it appears yellowish).
6. Add another ¼ cup of milk to the water and stir it.
7. Repeat step 5. (Now the beam of light should look even more blue from the side and more yellow, maybe even orange, from the end)

8. Add the rest of the milk to the water and stir the mixture.
9. Repeat the flashlight process. The beam should continue to look even bluer, and from the end, it should look very orange.

What's happening?

Light usually travels in straight lines, unless it encounters the edges of something. When the beam of your flashlight traveled through water, you couldn't see the beam from the side because the water is uniform, and the beam travels in a straight line.

When you added milk to the water, you also added many tiny particles. Milk contains tiny particles of protein and fat, and these particles scattered the light and made the beam of the flashlight visible from the side. Different colors of light are scattered differently. For example, blue light is scattered much more than orange or red. Looking at the side of the beam, you see the scattered blue light, and it looks blue. When you look directly at the beam of the flashlight, it looks orange or red because those colors are scattered less and they travel in a straight line from the flashlight.

The sunlight you see during the day is scattered by dust particles in the atmosphere in the same way as the light from the flashlight is scattered by particles in milk in this experiment. Looking at the sky is like looking at the flashlight beam from the side: you're looking at scattered light that is blue. When you look at the setting sun, it's like

looking directly into the beam from the flashlight: you're
seeing the light that isn't scattered, namely orange and red.

105)   COOL LIGHT

Have you ever heard of the word, *Chemiluminescence*? It means cool light! Learn how it works with this "cool" experiment.

Materials you'll need:

- Glowstick (available at most camping supply, hardware, or sporting goods stores)
- Dark room
- Glass of ice water
- Freezer
- Pen and paper (to describe the changes you see)

What to do:

1. Open the wrapper and remove the Glowstick.
2. Write down a description of the Glowstick. What does it look like? Is anything inside of it?
3. Follow the directions on the wrapper to activate the Glowstick (bend and shake it).
4. Observe the Glowstick in a darkened room. Describe its appearance.
5. Immerse the Glowstick in a glass of ice water for 5 minutes. What happens to the glow?
6. Put the glowing Glowstick in the freezer for at least 24 hours. Does the Glowstick continue to glow while it is in the freezer?
7. Remove the Glowstick from the freezer and allow it to warm to room temperature.

8. Throughout the day, check on your Glowstick and record any changes.

What's happening?

Many chemical reactions produce both light and heat, like a burning candle or light bulbs. It is less common for a chemical reaction to produce light without heat. When it does, this is called a *chemiluminescent reaction*, or cool light. Some living organisms, like fireflies produce light without heat by a chemiluminescent reaction. This experiment shows you that like all chemical reactions, the chemiluminescent reaction is slower at lower temperatures.

106)   GLOW IN THE DARK EXPERIMENT (PART 1)

Materials you'll need:

- A black light
- Petroleum Jelly
- Piece of Paper
- Tissue or cloth

What to do:

1. Dip your finger into the petroleum jelly, and write a short message on the piece of paper.
2. Wipe any remaining jelly off your finger with the tissue or cloth.
3. Turn off the room lights and turn on the black light.
4. Can you see the message?

What's happening?

The reason black lights are called "black lights" is because they give off very little light that our eyes can see. Instead, the black light creates a type of ultraviolet (UV) light, which illuminates things called phosphors, which absorbs radiation and releases it as light we can see. Petroleum jelly contains these phosphors, so the glow you see is from the phosphors absorbing the invisible ultraviolet radiation and emitting visible light.

107) GLOW IN THE DARK EXPERIMENT (PART 2)

Make your hands glow with Part 2 of this experiment!

Materials you'll need:

- A black light
- Petroleum Jelly
- Latex gloves
- A helper (to turn on the black light for you)

What to do:

1. Put the gloves on your hands.
2. Reach into the jar of petroleum jelly and scoop out enough to fully cover both hands. (Rub the jelly well over both hands)
3. Ask your helper to turn off the lights in the room, and to turn on the black light.
4. Hold your hand under the black light.

What's happening?

Like Part 1 of this experiment, the phosphors in the petroleum jelly are absorbing the black light's UV radiation and emitting visible light, but this time it's your hands that are glowing! Cool, huh? Can you think of a way you could use this trick to tell scary stories?

108)   POTATO BATTERY

This cool experiment will show you how chemical energy can become electrical energy.

Materials you'll need:

- 1 potato
- 1 plate
- 2 pennies
- 2 nails
- 3 pieces of insulated copper wire (about 8 inches long, each with 2 inches of the insulation stripped off one end)
- Digital clock (with attachments for wires).

What to do:

1. First, cut your potato in half and put the two halves on a plate so they are standing up on their flat ends.
2. Wrap the end of one piece of wire around a nail, and then wrap the end of another piece of wire around a penny.
3. Stick the nail and penny into one half of the potato, but make sure they're not touching each other.
4. Wrap another piece of wire around the second penny and put it into the other half of the potato.
5. Put the second nail into the potato as well, but don't wrap any wire around it.

6. Okay, now take the wire attached to the penny on the
   first potato, and connect it to the nail (that has no wire)
   on the second potato.
7. Now touch all the free ends of the potato wires to the
   wires coming out of the digital clock.
8. Does it work? (You may have to try connecting the
   wires to the clock in different ways to get it to work).

What's happening?

When the nails and copper wires are put together, they
generate heat (also known as energy). With the addition of
the potato, the properties of the nail and copper stay
separated, which causes the chemical reaction (the heat) to
become electric energy which can be used to power small
devices like your digital clock.

109)   FLAME RELIGHT EXPERIMENT

This experiment will teach you more about what fire needs to burn (and it'll teach you a cool trick!). Since this experiment involves fire, be sure to get help from an adult.

Materials you'll need:

- Yeast
- Baking soda
- White vinegar
- Hydrogen peroxide
- Popsicle stick
- Two tall drinking glasses (or graduated cylinders, if you have them!)
- Lighter
- Measuring spoons
- Safety glasses

What to do:

1. Add 1 teaspoon of baking soda to both glasses (or graduated cylinders)
2. Add 1 teaspoon of yeast to both of the both.
3. Spin and twirl the glasses mix up the baking soda and yeast.
4. In one glass, add a generous pour of hydrogen peroxide.
5. In the other glass, add a generous pour of white vinegar.
6. With the help of an adult, use a lighter to ignite a Popsicle stick. Wait until you have a strong, consistent flame before you continue.

7. Stick the lit end of the stick down into the glass that contains vinegar. The flame should extinguish, even without touching the liquid.
8. Now, take the still-smoking stick, and hold it down into the glass that contains hydrogen peroxide without touching the liquid. The flame should relight!
9. Repeat going back and forth between the glasses as many times as you can.

What's happening?

Fire needs oxygen, fuel, and sufficient heat to ignite and stay afire. If any of these elements is removed, the fire goes out. When you stick the flaming stick into the first glass, it goes out because the carbon dioxide (CO_2) removes oxygen from the equation. The carbon dioxide is formed from combining the base of baking soda (sodium bicarbonate) with the acid of vinegar (acetic acid). This combination creates *carbonic acid* which decomposes into carbon dioxide (CO_2) and water (H_2O).   The bubbling in the container is carbon dioxide gas. When you dip the flaming stick into the cylinder, you're exposing the flame to concentrated CO_2 gas, which removes the oxygen and puts out the flame.

When you place the stick into the second glass (with the hydrogen peroxide), the ember glows until the flame reignites. This is because a high concentration of oxygen (O_2) is reintroduced, making it light afire again. When hydrogen peroxide (H_2O_2) is combine with yeast, the yeast acts as a catalyst for hydrogen peroxide decomposition into water (H_2O) and oxygen (O_2). So oxygen is created at a

much faster rate and by placing the partially glowing ember
the glass, you reignite the flame.

XI.   INERTIA, FORCE, AND FRICTION

110)   NO-SPILL BUCKET SPINNING

Learn about centripetal force with this fun science experiment! Be prepared to get a little wet in the beginning, but soon you will be able to do this experiment without spilling a drop!

Materials you'll need:

- A bucket with a strong and sturdy handle
- Water
- An open area outside, like a lawn or park.

What to do:

1. Fill the bucket with water until it is half full.
2. Stand in an area away from people and anything else that could get in the way.
3. Hold the bucket by the handle with a straight arm. Start spinning in circles, and raise the bucket towards the sky and back to the ground (make sure to spin it fast enough to keep the water inside the bucket).
4. Stop spinning before your arm gets tired, and carefully bring the bucket to rest on the ground.

What's happening?

This experiment involves called *centripetal force*, which is the force that acts on an object moving in a circular path. The force directs everything towards the center around which it is moving, which is why the water doesn't fly out of the bucket and onto you! *Centripetal* force can be seen all over the earth—this is how roller coasters don't fly off the rails, and how satellites stay in orbit around a planet.

111) INERTIA HIGH DIVE

Materials you'll need:
- A penny (or other small coin)
- A piece of card stock or stiff paper
- A film canister (or baby food jar, or other small container)
- Pencil or pen
- Scissors

What to do:

1. Cut the cardstock paper into a long strip about 1 inch wide and form it into a hoop.
2. Fill the film canister with water and place on a level surface.
3. Place the hoop on the canister and balance the penny on the top of the hoop.
4. Now place the pencil through the center of the hoop, and quickly fling the hoop off to the side.

What's happening?

By moving the pencil, you pass on the energy of movement to the hoop, which makes it fly out of the way quickly. You would think this would take the penny with it, but there wasn't enough *friction* to affect the penny on top. So the penny stays in place, directly over the canister, and immediately *gravity* took over, and pulled the coin straight down into your container.

112)   INERTIA TOWER

This simple science experiment is a great demonstration in inertia, friction, and movement.

Materials you'll need:

- Coins of the same size
- Butter knife

What to do:

1. Stack the coins into an even and straight tower.
2. Use the butter knife to swipe a coin out from the bottom of the tower.
3. See how many coins you can swipe out before the tower falls.

What's happening?

According to *Newton's First Law of Motion*, an object in motion (or at rest) tends to stay in motion (or at rest). This is also known as *inertia*. This means that the balanced coin tower wants to stay in its stacked position, where they are stacked. However, applying an *outside force* (in this case, your knife) can cause the coins change position, or fall over. The amount of *force* you apply to the coin is enough that tower drops into the spot that it was before.

113) WATERFALL INERTIA

Materials you'll need:

- Long strand of beads
- Drinking glass or other large container
- Tape (any kind)

What to do:

1. In order for this experiment to work, you must put the beads in the glass correctly to prevent tangling. To do so, find one end of the string of beads and mark it with a piece of tape. Now, place the end into the container and continue to feed the beads into it (being careful not to tangle the string) until the whole strand is in the container.
2. Mark the other end of the string with a piece of tape.
3. Hold the container up high with one hand grasp the end of the string of beads (the taped bit).
4. Now, using a fast-pulling motion, give a quick tug to the string of beads and let go (while still holding the container high).
5. The rest of the string of beads should instantly "climb over" the side of the container and land on the floor.

What's happening?

The weight of the first small section of beads that you pull out, should be enough to pull the rest of the beads out of the container. Holding the container high is an example of *potential energy* (stored energy). When you gave a quick yank to the beads, you turn *potential energy* into *kinetic energy* (moving energy). *Inertia* keeps the strong of beads moving towards the floor.

114)   DOLLAR BOTTLE

This is a cool trick you can show your friends while learning more about friction!

Materials you'll need:

- Dollar Bill
- Glass Bottle

What to do:

1. Place a dollar bill flat on a smooth surface.
2. Balance a bottle (upside down on its mouth) right over the center of the bill.
3. Try pulling the dollar bill out from under the bottle. Try a few times.
4. Now, carefully roll the end of one side of the dollar bill towards the bottle.
5. Once you have rolled it all the way to the mouth of the bottle, keep rolling with care as the bottle gets nudged towards the opposite edge.

What's happening?

Inertia and *friction* allow you to move the bill without knocking over the bottle. If you remember from our **Inertia Tower Experiment,** *inertia* means that the balanced bottle wants to stay in its spot, but when you apply an *outside force* by attempting to remove the dollar bill underneath the bottle, it wants to fall over. This is because of the *friction* between the bill and bottle—there's so much friction that

the bill pulls the bottom of the bottle with it. By rolling up
the dollar bill to the edge of the bottle, you overcome the
friction, without using so much movement to tip it over.

115)   CENTRIPETAL FORCE BOARD

This experiment is a more involved version of our **No-Spill Bucket Experiment**, with even more impressive results! You'll need some tool for this experiment, so be sure to get an adult to help.

Materials you'll need:

- Rope
- Water
- Three plastic cups
- Thin square board
- Square sheet of rubber
- Drill
- Heavy duty glue

What to do:

1. Get your adult helper, and drill a hole large enough for the rope in each corner of the board.
2. Glue the sheet of rubber to one side of the board.
3. Cut your piece of rope into two even lengths, and tie the center of them together.
4. Pull one strand of rope through each hole and tie the ends.
5. Hold up the board to make sure the ropes are even—the board should lie flat when you hold the ropes by the tied center.
6. Now place the 3 plastic cups in the center of the board and fill them with water.

7. Slowly swing the board, and eventually swing it in a
    complete circle.
8. When you're done swinging it, bring it to a slow stop.

What's happening?

Remember *Newton's First Law of Motion*? The one that
says objects in motion tend to remain in motion? In this
case, when you start to swing board the board in a circle,
you start the water moving along the curve of that same
circle. *Centripetal force* takes over at this point, which is
the force that keeps it always turning toward the center of
the circle.

116) FORCEFUL PENNY

Another fun experiment showcasing *Newton's First Law of Motion*. This involves bending wire, so ask an adult for help!

Materials you'll need:

- Wire hanger
- penny

What to do:

1. With the help of an adult, bend the hook part of your wire hanger until the end is pointed back in the opposite direction.
2. Stretch the triangle part of the hanger to open it up. It should look similar to a diamond.
3. Balance the penny on the hooked end of the hanger. Be patient! It may take a few tries…
4. Begin to swing the hanger back and forth, starting small and gradually increasing the swing until you can spin it in a full circle. Again, it may take a few times, but you'll get it!

What's happening?

Newton's law says that the penny will continue moving along the curve of the circle, but a *force* is required to keep it always turning toward the center of the circle. *Force* the swinging of the hanger, which causes your penny to accelerate.

117) FRICTION NOTEBOOKS

Materials you'll need:

- 2 notebooks
- Another person

What to do:

1. Place the notebooks on a flat surface with the bindings facing inward
2. Make sure the covers completely overlap with each other
3. Alternate pages from each notebook placing one over the last, continuing until the notebooks are entirely intertwined.
4. Have a friend grab one notebook by the binding and you grab the other (from the binding also) and both of you pull to see if you can break them apart.

What's happening?

Friction is at work here, which is the force that opposes motion when two surfaces are in contact. The amount of friction between sheets of paper is pretty minimal, but when you take a lot of paper and interweave them together like you did here, you wind up with so much friction that you can't break it.

XII. EXPERIMENTS WITH CANDY

118)   MAKE A GUMMY FROG SWIM!

In this experiment you will learn how absorption can change the size of things and see if you can make your gummy frog swim!

Materials you'll need:

- A Haribo gummy frog (or any other gummy candy with a marshmallow coating on the bottom)
- Clear bowl of water

What to do:

1. Drop the frog into your bowl of water.
2. Wait a couple of days... What happens to the frog? What happens to the marshmallow?

What's happening?

The frog is porous, so as it absorbs water, several things start to happen. It expands and its density (weight and volume) changes. Its density starts to be more and more like the water surrounding it. As it grows, it also stretches out the marshmallow, which expands the amount of air bubbles. (If you look closely, you should be able to see them.) Its weight also changes, because its heavy sugar is dissolving into the surrounding water. The combination of a change in the frog's density along with the air bubbles underneath the frog, is enough to gently lift it of the bottom of the bowl.

119)   SALTWATER GUMMY CANDY EXPERIMENT

Use this saltwater experiment to learn about osmosis!

Materials you'll need:

- 4 gummy worms (or any other gummy candy)
- 3 small clear bowls with 1 cup of water each
- Sugar
- Salt

What to do:

1. In one bowl of water, add a spoonful of sugar.
2. In the second bowl of water, add a spoonful of salt.
3. Place a gummy worm in each bowl. Set the other gummy candies aside for comparison.
4. After several hours, check the gummy worms to see what size they are. Which ones have changed the most?

What's happening?

This process is called osmosis. Your bowl of fresh water is considered a "dilute" solution. This means it is not very strong. When you add salt to it, you change the water from being a "dilute solution" to "concentrated solution." The more salt, the more concentrated the solution becomes. Osmosis is the process where water tries to pass from a dilute solution into a concentrated one. It is trying to balance the two solutions.

This experiment lets you see osmosis in action. When you put a gummy worm into fresh water, it flows into the gummy worm, diluting the sugar-gelatin mix it is made of (a concentrated solution). The gummi worm expands. But if you put a gummi worm in salt water, the salt water is already concentrated. Not as much water is needed to dilute the gummy worm, because the salt water and the sugar-gelatin mix in the gummy candy are already balanced. Not as much water passes into the gummy worm, and it will expand less. In fact, if your solution is super concentrated (a lot of salt), the gummy candy might not expand at all.

120)   MELTING CHOCOLATE EXPERIMENT

What temperature makes chocolate change from a solid to a liquid? Try this melting chocolate experiment and find out!

Materials you'll need:

- Small chocolate pieces of the same size
- Paper plates
- Pen and paper to record your results

What to do:

1. Place one piece of chocolate on a paper plate and put it outside in the shade. Write down the time.
2. Wait until it melts and write down how long it took (or if it isn't hot enough to melt, then record how soft it was after 10 minutes).
3. Repeat the process, but this time put the chocolate outside directly in the sun. Record your results in the same way.
4. You can try this with many different places and temperatures! Try hot water or even your mouth.
5. Compare your final results--in what conditions did the chocolate melt the fastest? The slowest? To explore further, use a thermometer to record the temperatures of the locations you used and compare.

What's happening?

Your chocolate pieces are going through a physical change, from a solid to a liquid (or somewhere in between). You can also reverse the process by putting the melted chocolate into a fridge or freezer where it will undergo another physical change, this time from a liquid back to a solid. If you used your mouth to melt chocolate, what does this say about our body temperature?

MORE: Compare white chocolate and dark chocolate—do they melt at the same temperature? Try placing a sheet of aluminum foil under a piece of chocolate in the sun—does melting speed up or slow down?

121)   LIGHTNING CANDY

Materials you'll need:

- A dark room (a bathroom works great)
- Your mouth
- A Wintergreen Life Saver candy
- A mirror

What to do:

1. Turn out the lights and let your eyes to become accustomed to the dark.
2. Look into the mirror and place a wintergreen Life Saver candy into your mouth
3. While looking at the mirror, chew the candy with your mouth open.
4. What do you see?

What's happening?

When the sugar in the candy is cracked, *charge separation* occurs. The sugar in the candy has a crystalline structure that causes *electrons* (particles with a negative charge) to be torn away from the nucleus of their atoms. This separates the *electrons* from the *protons* (particles with a positive charge), which creates an electrical field strong enough to take electrons from molecules in the air (a process called *ionization)* which, in turn, creates plasma ions. These plasma ions then run into into other molecules and transfer their energy which emits *photons* of light. Those are the flashes you observed. This crazy phenomenon is called *triboluminescence*, and I know it's a lot of information, but isn't it cool?!

122)   TRANSLUCENT TAFFY

This experiment uses an oven, so be sure to get an adult's help with it.

Materials you'll need:

- Saltwater taffy, Laffy Taffy, or other kind of Taffy.
- Foil-lined baking sheet
- Oven

What to do:

1. With an adult's help, preheat the oven to 300°F.
2. Place the unwrapped candy on the baking sheet.
3. Heat the candy in the oven for up to 30 minutes, checking every few minutes.
4. What does the candy look like?

What's happening?

Taffy is made in a machine that has metal arms which stretch and fold the candy over and over. As this happens, tiny air bubbles get trapped in the candy—that's what makes it soft and chewy. When you melt taffy in the oven, some of the air bubbles escape, the taffy becomes translucent again.

123)   MAKE ROCK CANDY AT HOME

This experiment deals with very hot liquids, so ask an adult for help.

Materials you'll need:

- A wooden skewer (you can also use a clean wooden chopstick)
- A clothespin
- 1 cup of water
- 2-3 cups of sugar
- A tall narrow glass or jar

What to do:

1. Clip the wooden skewer into the clothespin so that it hangs down inside the glass and is about 1 inch from the bottom of the glass
2. Remove the skewer and clothespin and put them aside for now.
3. With the help of an adult boil your water.
4. Pour about 1/4 cup of sugar into the boiling water, stirring until it dissolves.
5. Keep adding more and more sugar, each time stirring it until it dissolves, until no more will dissolve. (Be patient)
6. Once no more sugar dissolves, remove it from heat and allow it to cool for 20 minutes.
7. Add food coloring to the sugar water.

8. Now have your adult helper pour the sugar solution into
    the jar almost to the top.
9. Submerge the skewer back into the glass making sure
    that it is hanging straight down the middle without
    touching the sides.
10. Allow the jar to fully cool and put it someplace where it
    will not be disturbed.
11. Now just wait. The sugar crystals will grow over the
    next 3-7 days.

What's happening?

When you mixed the water and sugar you made a *super
saturated solution*. This means that the water could only
hold the sugar if both were very hot. As the water cools the
sugar "comes out" of the solution back into sugar crystals
on your skewer. The skewer acts as a "seed" that the sugar
crystals can grow on, eventually giving you a tasty treat!

124)   TASTE-TESTING WITH PH

In this experiment, you get to taste-test acidity and then double-check your taste buds with a pH strip. Now that's science!

Materials you'll need:

- Sour candies (such as Sour Skittles, Lemonheads, Sour Patch Kids, etc.)
- Small bowls
- pH indicator paper
- Warm water

What to do:

1. Put a few pieces of candy in each bowl. (Use a different bowl for each kind of candy.)
2. Pour warm water into each bowl until the candy is covered.
3. Let the sour part of the candy dissolve.
4. Taste-test a spoonful of water from each bowl. How sour is it?
5. Arrange the bowls by order of how sour they taste to you.
6. Now, test each bowl of water with a pH indicator strip.
7. Do the results match the results of your tasting?

What's happening?

Sour taste is caused by acid, so the more sour your candy is, the more acid it contains. If your taste test doesn't exactly match the pH paper test, it could be because some flavors used in candy may cover up more acidic tastes.

125) GOBSTOPPER SURPRISE

This colorful science experiment may surprise you.

Materials you'll need:
- 4 Gobstopper candies
- Petri dish (or shallow dish)
- Water

What to do:

1. Pour water into your dish until the bottom is covered with a thin layer.
2. Place your Gobstoppers equal distances from each other along the edge of the dish. One at the top, one at the bottom, and the last two on opposite sides.
3. What do the colors do?

What's happening?

The surprise in this experiment is that the colors of the Gobstoppers don't mix, but instead run into each other and stop as if they each hit a color wall. This is because Gobstopper candy contains *carnauba wax,* which forms a shield that prevents the water soluble dyes from mixing. If you remember, Gobstoppers have many different layers of colors, the *caranauba wax* is responsible for their separation. If you keep watching the pools of color should change as the Gobstopper slowly dissolves.

126)   LEARN ABOUT DENSITY LAYERS WITH POP ROCKS

This fun experiment helps you learn about different densities, with candy!

Materials you'll need:

- 3 packages of Pop Rocks (two packages of one color, one package of a different color)
- Drinking glass filled with water

What to do:

1. Pour 2 packages of the same color Pop Rocks into the glass of water.
2. Wait several minutes until the Pop Rocks dissolve.
3. Pour the other package of colored Pop Rocks into the water.
4. What happens?

What's happening?

As we have learned in some of our other experiments, a less-dense liquid will float on top of a denser liquid. When you dissolve the first two packages of Pop Rocks, you create a dense *sugar solution* at the bottom of the cup. When you add the third package, the Pop Rocks float on top of that dense solution for a little bit before they dissolve into a cool stripe of color.

127)   WARTY LICORICE

Can you make a smooth piece of licorice grow warts? This experiment uses a microwave so be sure to ask an adult for help.

Materials you'll need:

- Twizzlers licorice twists
- A plate
- Microwave

What to do:

1. Place the licorice on the plate.
2. Start microwaving on low heat, and check it every 30 seconds (so it doesn't burn).
3. What happens?

What's happening?

Licorice contains a little water. The warts are caused by water making tiny pockets of steam when it's heated. The licorice doesn't melt because it contains flour (which doesn't melt).

128)   INCREDIBLE GROWING AND SHRINKING PEEPS

This yummy experiment uses a vacuum sealer, so be sure to get the help of an adult.

Materials you'll need:

- A kitchen vacuum sealer
- A vacuum sealer storage container
- Marshmallow Peeps of various sizes (or any kind of marshmallow will do)

What to do:

1. Fill the storage container jar with Peeps and place the lid on top.
2. With the help of an adult, attach the vacuum hose from the sealer to the container.
3. Turn on the sealer (follow the directions!)
4. What happens to the marshmallows?
5. Now release the vacuum seal—what happens?

What's happening?

The job of the vacuum sealer is to remove the air from inside the container. Normally, molecules of air in the atmosphere (*atmospheric pressure*) are pushing on the outside of the Peep marshmallow. When the vacuum sealer takes away that air that was pushing on it, the air trapped inside the marshmallow *expands* (pushes out) which allows the peep to get bigger. The marshmallows shrink back to normal size when the seal is broken and air rushes back into the container.

129) CANDY CHROMATOGRAPHY

Ever wondered why candies are different colors? Many candies contain colored dyes. We can answer this by dissolving the dyes out of the candies and separating them using a method called chromatography.

Materials you'll need:

- M&Ms or Skittles (1 of each color)
- Coffee filter paper
- A tall glass
- Water
- salt
- a pencil
- scissors
- ruler
- 6 toothpicks
- aluminum foil
- an empty 2-liter plastic bottle with cap

What to do:

1. Cut the coffee filter paper into a 3 inch by 3 inch square. Draw a line with the pencil about ½ inch from one edge of the paper. Make six dots with the pencil equally spaced along the line, leaving about ¼ inch between the first and last dots and the edge of the paper. Below the line, use the pencil to label each dot for the different colors of candy that you have.
2. Now, make solutions of the colors in each candy. Take a large piece of aluminum foil and lay it on a table.

3. Place six drops of water spaced evenly along the foil. Place one color of candy on each drop. Wait about a minute until some of the color is dissolved from the candies into the water. Remove and throw away the candy.
4. Now take a toothpick and wet it in one of the colored solutions and lightly touch it to the corresponding labeled dot on your coffee filter. Use a light touch, so that the dot of color stays small. Then repeat, using a different toothpick for each color.
5. Wait for each spot to dry, and then repeat the whole process again three more times to get more color on each spot.
6. When the paper is dry, fold it in half so that it stands up on its own, with the fold standing vertically and the dots on the bottom.
7. Now, make a *developing solution*. Add ⅛ teaspoon of salt and three cups of water into your clean and empty 2-liter plastic bottle. Put the cap on tightly and shake it until all of the salt is dissolved in the water.
8. Now pour the salt solution into the tall glass to a ¼ inch depth (it should be low enough so that when you put the filter paper in, the dots will initially be above the water level).
9. Hold the filter paper with the dots at the bottom and set it in the glass with the salt solution.
10. As the salt solution climbs up the filter paper, what do you begin to see?
11. When the salt solution is about ½ inch from the top edge of the paper, remove the paper from the solution. Lay the paper on a flat surface to dry.

What's happening?

This experimental process is called *chromatography*. The color spots climb up the paper along with the salt solution (due to *capillary action)*, and some colors start to separate into different bands. The colors of some candies are made from more than one dye, and the colors that are mixtures separate as the bands move up the paper. The dyes separate because some dyes stick more to the paper while other dyes are more soluble in the salt solution.

Compare the spots from the different candies, noting similarities and differences. Which candies contained mixtures of dyes?

MORE: Try different candies, and you even use chromatography to separate colors from colored markers, food coloring, and Kool-Aid.

130) COLORFUL SUGAR GLASS

This scientific experiment teaches you about crystallization AND makes a delicious treat. You need a stove for this one, so ask an adult for help!

Materials you'll need:

- Granulated sugar
- Saucepan
- Food coloring
- Cream of tartar
- Candy thermometer
- Light corn syrup
- Measuring utensils
- Aluminum foil pan
- Cooking Spray
- Water

What to do:

1. Measure your ingredients:
- 1 ¾ cups sugar
- 1 cup water
- ½ cup corn syrup
- 1/8 teaspoon cream of tartar
2. Combine all the ingredients in your saucepan.
3. With your adult helper, slowly heat the mixture to a low boil while stirring the whole time. This is a slow heat, so be sure not to heat it too quickly.

4. Keep the mixture at a low boil and use your candy thermometer to test it occasionally. You want to keep it boiling until it reaches 300ºF.
5. While waiting, prepare your foil pan by spraying it lightly with cooking spray.
6. When the mixture reaches 300ºF, CAREFULLY pour the mixture into the foil pan (be sure you have help with this!)
7. Add whatever food coloring you want to the pan and spread the colors with a knife or fork.
8. Let the mixture cool until hardened.
9. Remove your "Stained Glass" from the foil pan and admire your beautiful handiwork!

What's happening?

"Sugar glass" is created when you dissolve sugar into water and heat it to 300ºF. By adding corn syrup, you prevent the sugar from recrystallizing. This happens because corn syrup prevents the sugar molecules in the mixture from bonding (which forms crystals). By adding cream of tartar you separate the original, complex sugar crystals into *glucose* and *fructose*, which are known as *simple sugars.* And by adding in food coloring you create a lovely "stained glass" look.

XIII. AIR PRESSURE & WEATHER EXPERIMENTS

131) BUILD AND TEST A PARACHUTE

Make an awesome parachute and learn about air resistance!

Materials you'll need:

- A plastic bag or other light material
- Scissors
- String
- A small object to act as the weight

What to do:

1. Cut out a large square from your plastic bag
2. Trim the edges so it looks like an octagon (an eight-sided shape).
3. Near the edge of each side, cut a small hole, for a total of 8 holes.
4. Cut your string into 8 equal sized pieces.
5. Attach one piece of string to each of the holes.
6. Gather the loose ends of the strings together and tie them to your weight.
7. Find a high spot, or use a chair, to drop your parachute.

What's happening?

When you drop the parachute, the weight pulls down on the strings and opens up a large surface area of the material. This open area uses air resistance to slow it down. The larger the surface area the more air resistance, and the slower your parachute will fall.

MORE: To make the parachute fall in a straighter line, cut a small hole in the middle of the material. This will let air slowly pass through it, rather than spilling out over one side, giving it a smoother ride down.

132)   FANTASTIC FLYING TRASH BAG

Do you think you can make a trash bag fly? Try this fun experiment to find out!

Materials you'll need:

- 1 large black garbage bag
- Strong tape, such as duct tape
- Long piece of string
- Hair Dryer
- An open area outside
- A sunny day (with no wind)

What to do:

1. Hold the mouth of a trash bag in one hand.
2. Use your hair dryer to blow hot air into the bag until it's mostly full.
3. Seal the mouth of the bag closed with duct tape.
4. Now tie one side of your string around the tape and hold the extra bit with your hand.
5. Take the bag out into the sun. What happens?

What's happening?

The fantastic flying trash bag should slowly rise into the air. Because the trash bag is black, it absorbs heat from the sun, which makes the air inside the trash bag *expand* (get bigger) and become lighter. When the bag and the air inside it are lighter than the surrounding air outside, it will start to rise.

133)   MAKE A BALLOON ROCKET

Materials you'll need:

- 1 balloon
- 1 long piece of kite string (10-15 feet long)
- 1 plastic straw
- Tape

What to do:

1. Tie one end of the string to a chair, door knob, or other support.
2. Put the other end of the string through the straw.
3. Pull the string tight and tie it to another support in the room.
4. Blow up the balloon (but don't tie it.) Pinch the end of the balloon and tape the balloon to the straw.
5. Let go!

What's happening?

As the air rushes out of the balloon, it creates a forward motion called *thrust.* Thrust is a pushing force created by energy. In the balloon experiment, our thrust comes from the energy of the balloon forcing the air out. Different sizes and shapes of balloon will create more or less thrust

134)   SMOKE CANNON

With some adult supervision, learn about vortexes while you make awesome smoke rings!

Materials you'll need:

- Empty plastic coffee tub
- Utility knife
- Smoke bomb
- Lighter
- Space outside

What to do:

1. With adult help, cut a circle out of the bottom of the coffee tub using the utility knife.
2. Light the smoke bomb outside and let it start smoking.
3. Quickly and carefully put the smoke bomb in the coffee tub and put the lid on.
4. Point the hole of the coffee tub away from you and tap the lid of it to create smoke rings!

What's happening?

The air ring that shoots out of the cannon is *a flat vortex of air*. The air exiting the cannon is going faster in the middle of the hole than the air on around the edge of the hole, which creates…you guessed it, a vortex. The colored smoke from the smoke bomb highlights how air occupies space. When playing with this cannon, have fun, use different colors, but remember to never point it at a person or animal.

135)   WATER CANDLE

This experiment shows how a candle can suck water! This experiment involves matches, so be sure to ask an adult for help!

Materials you'll need:
- Candle
- Water
- Saucer or small dish.
- Matches (ask for help!)
- Tall drinking glass

What to do:

1. Place the candle upright in the middle of the saucer.
2. Fill the saucer with water.
3. Light the candle.
4. Place the glass over the lit candle.
5. What happens when the flame goes out?

What's happening?

When the candle is burning inside the glass, the heat from its flame makes the air expand so some of the air escapes from the glass. Once the candle has used all the oxygen in the glass the fire goes out and the inside of the glass starts to cool, which drops the air pressure. The water outside the glass is sucked into the low pressure area in the glass by the higher air pressure outside.

136) BUILD A HOOP GLIDER

Materials you'll need:

- A regular plastic drinking straw
- 3 x 5 inch index card (or other stiff paper)
- Tape
- Scissors

What to do:

1. Cut the index card into 3 separate pieces that measure 1 in x 5 in.
2. Take 2 of the pieces of paper and tape them together into a hoop. (If you overlap the pieces by half an inch, it will keep a nice round shape when secured with tape).
3. Use the last strip of paper to make a smaller hoop.
4. Stick the tip straw through the hoops and tape them in place, with the smaller hoop closest to the edge.
5. Now hold the straw in the middle with the hoops on top and it like a dart.
6. What happens?

What's happening?

The two different hoops sizes keep the contraption balanced as it flies. The big hoop creates *drag* (air resistance) which helps keep the straw level, and the smaller hoop in at the front keeps it from turning off course.

137)   THE COLLAPSING CAN

In this experiment, you'll use air pressure to crush an aluminum can. This experiment uses the stovetop, so be sure to ask an adult for help!

Materials you'll need:

- 1 empty aluminum can
- A deep saucepan
- A pair of kitchen tongs
- stovetop

What to do:

1. Fill the saucepan with cold water.
2. Put 1 Tablespoon of water into the empty can.
3. With help from an adult, place the can on the kitchen stove to boil the water inside.
4. When the water boils, you'll see a cloud of condensed vapor escape from the mouth in the can. Allow the water to boil for about 30 seconds.
5. Using the tongs, grasp the can and quickly dip it into the cold water bath.

What's happening?

When you boiled the water, the vapor pushed all the air out of the can. When the can was filled with water vapor, you cooled it suddenly when you put it in the cold water bath. This made the water vapor inside the can condense, which created a *partial vacuum*. The *low pressure* of the partial vacuum inside the can allowed the air pressure outside of the can to crush it.

138)   BOTTLE BALLOON MYSTERY

How easy do you think it is to blow up a balloon while it's inside a bottle?

Materials you'll need:

- a balloon
- a large plastic bottle (2-liter works best)
- thumbtack
- nail

What to do:

1. Place your balloon inside the bottle, but stretch the neck over the bottle opening.
2. Try to blow the balloon up while it's in the bottle. What happens?
3. Take out the balloon, and use the thumbtack to punch a hole in the bottom of the bottle.
4. Use a nail to make the hole larger, until it is at least ¼ inch big in diameter.
5. Place the balloon back into the top of the bottle and stretch the neck over the bottle opening.
6. Try blowing it up again. What happens?

What's happening?

To blow up any balloon, you force air inside it. But when you try to blow up a balloon while it's inside an empty bottle, you can't because the bottle is actually already filled with air. Our lungs can't exert enough pressure to force the air out of the bottle. However, once you poke a hole in the bottom of the bottle, some of the air can escape, which allows you to blow more air into the balloon!

139)   BUILD A MINI HOVERCRAFT

Materials you'll need:
- An old CD or DVD disc
- A 9" balloon
- A pop-top cap from a liquid soap bottle or a water bottle
- A hot glue gun

What to do:

1. Cover the center hole of the CD with a piece of tape and poke about 6 holes in the tape with a push-pin or small nail (this will slow down the flow of air and allow your hovercraft to hover longer.
2. Use the hot glue gun to glue the cap to the center of the CD or DVD disc. Create a good seal to keep air from escaping.
3. Blow up the balloon all the way and pinch the neck of it.
4. Make sure the pop-top is closed and fit the neck of the balloon over the pop-up portion of the cap.
5. When you are ready see your hovercraft, put it on a smooth surface and simply pop the top open

What's happening?

The air flow created by the balloon causes a cushion of moving air between the disc and the surface. This lifts the CD and reduces the friction which allows the disc to hover freely.

140)   HOVERING PING-PONG BALL

Put your coordination skills to the test while learning about gravity and air pressure in this experiment.

Materials you'll need:

- 1 Ping-Pong balls
- A hair dryer

What to do:

1. Plug in your hair dryer and turn it on.
2. Put it on its highest setting and point it straight up, towards the ceiling.
3. Place your ping pong ball above the hair dryer and watch.

What's happening?

Your ping pong ball should float gently above the hair dryer. This happens because the air from the hair dryer pushes the ping pong ball upwards. When its upward force equals the force of gravity pushing down on it, the ball bounces around gently and floats.

MORE: Try to float 2 or 3 ping pong balls as an extra challenge.

141)   PING PONG PRESSURE

Here's another fun ping pong experiment to teach you about air pressure and Bernoulli's Principle.

Materials you'll need:

- Modeling clay
- Hairdryer
- 2 plastic drinking straws
- Ping pong balls

What to do:

1. Place the end of a straw into a large piece of modeling clay. The straw should be able to stand upright in the clay.
2. Repeat with the second straw.
3. Mold two smaller pieces of clay around the bottoms of two ping pong balls.
4. Place the ping pong balls, clay side down, on top of each of the two upright straws.
5. Try to adjust the position the balls so they are as level as possible.
6. Plug the hairdryer in and turn it on.
7. Aim the hairdryer between the two ping pong balls. The ping pong balls should draw together.
8. Position the hairdryer differently to see if the ping pong balls still push together.

What's happening?

It may seem weird that as you point the air at the ping pong balls, they draw together instead of apart. This is an example of *Bernoulli's principle* which states that fast moving air creates an area of low pressure. The *low pressure* between the balls allows the *higher pressure* surrounding them to push them towards each other.

142)   HOLEY STRAW POTATO

Raw potatoes are hard to the touch. Do you think it is possible to stab one with a plastic drinking straw? Find out with this fun science experiment and learn about air pressure.

Materials you'll need:

- Some plastic drinking straws
- A raw potato

What to do:

1. Hold a plastic drinking straw by its sides (be careful *not* to cover the hole at the top).
2. Try to stab the potato quickly. What happens?
3. Repeat the experiment with a new straw, but this time cover the hole at the top with your thumb.
4. What happens?

What's happening?

During the first part of the experiment, you probably only pierced the potato a little bit (or not at all). With the second attempt, you should have been able to pierce the skin and push the straw much deeper into the potato.

Without covering the straw's hole in the first part, the air is pushed out of the straw and you are unable to stab the potato. By covering the top of the straw with your thumb, you trap the air inside. When you stab the straw into the potato, the trapped air in the straw is forced to compress. This contained *air pressure* makes the straw strong enough to pierce the potato.

143)   HUFF AND PUFF EXPERIMENT

Place a small item in the mouth of a bottle and attempt to blow the object into the bottle using a straw.

Materials you'll need:

- Small paper ball
- 1-liter bottle
- Plastic straw

What to do:

1. Create a small paper ball by bunching up a piece of paper. It should be able to loosely fit inside the mouth of the 1-liter bottle.
2. Put the plastic bottle on its side.
3. Place the paper ball in the mouth the bottle.
4. Point the straw towards the mouth of the bottle and try to blow the paper ball into the bottle.
5. You'll notice that the paper ball wiggles around before flying out of the bottle.
6. Replace the paper ball in the mouth of the bottle and try again.
7. Why won't it go into the bottle?

What's happening?

The answer to this puzzle is inside of the bottle. The bottle is full of air, **and** while you can't blow air or the piece of paper into the bottle, you are moving a lot of air along the sides of the bottle. So when the air blows past the mouth of the bottle, it creates an area of *low pressure* behind it. This low pressure help the paper ball hop out of the bottle's mouth!

144)   JUMPING JACK SODA CAN

This awesome science experiment is a simple demonstration of physics. Have fun!

Materials you'll need:

- Empty aluminum can
- 2 coffee mugs

What to do:

1. Place one mug in front of the other
2. Place the empty aluminum can in one of the mugs.
3. Blow air down between the soda can and mug.
4. What happens? (You may need to adjust the distance between the mugs)

What's happening?

The jump from one coffee mug to the other is cause by *air pressure.* When you blow air in between the can and the mug you make an area of *high pressure* in between the two surfaces—the mug and the bottom of the can. As the pressure between the two surfaces increases (inside the mug) the air pressure on top of the can stays the same. The difference in pressure pushes the empty can out. The harder you blow, the more rapidly the air pressure will build up, and the higher and further it will jump!

145) BERNOULLI'S WINDBAG

This experiment poses the question—how many breaths will it take to fill a 2-meter long bag? Learn about Bernoulli's Principle to help!

Materials you'll need:

- Purchase a long plastic bag in the shape of a tube. (You can find them in toy stores, but you can also make your own using a Diaper Genie refill bag, which is at most department stores)
- Another person

What to do:

1. Tie a knot in one end of the bag.
2. Have you or a friend blow up the bag, with one of you keeping track of the number of breaths it takes.
3. Now, let all of the air out of the bag.
4. Have your friend hold onto the closed end of the bag.
5. Now hold the open end of the bag approximately 10 inches away from your mouth.
6. Using only one breath, blow as hard as you can into the bag. (Remember to stay 10 inches away from the bag).
7. Quickly seal the bag with your hand so that none of the air escapes.

What's happening?

Bernoulli observed that a fast moving stream of air is surrounded by an area of low atmospheric pressure. The faster the stream of air moves, the more the air pressure drops around the moving air. When you blow into the bag, higher pressure air in the atmosphere forces its way into the area of low pressure created by the stream of air from you. The long bag inflates quickly with one breath because air from the atmosphere is drawn into the bag along with the air stream from your big breath.

146)   RAIN CLOUD IN A JAR

Learn about how clouds hold water with this colorful science experiment!

Materials you'll need:
- Water
- Food Coloring
- 2 clear drinking glasses or vases
- Shaving cream
- An eyedropper

What to do:

1. Fill your clear vessel with plain water.
2. Fill another glass with water and food coloring.
3. Squirt shaving cream on top of the PLAIN water (this is your "cloud").
4. Now slowly pour your colored water onto the shaving cream in the other vessel.
5. What do you see?

What's happening?

It should begin "raining" colored water as the shaving cream "cloud" gets too heavy and can't hold the water anymore. As you squeeze out more and more colored water, your "cloud" will soon release the excess below into the glass. This is just like how real clouds get too heavy with water and start raining.

147)   CLOUD IN A BOTTLE

Here's a quick and fun science experiment you can do—
make a cloud inside a bottle.

Materials you'll need:

- 1-liter plastic bottle
- warm water
- match

What to do:

1. Pour just enough warm water in the bottle to cover the bottom of the container.
2. Light the match and place the match head inside the bottle.
3. Allow the bottle to fill with smoke.
4. Cap the bottle.
5. Squeeze the bottle really hard a few times. When you release the bottle, you should see the cloud form. It may disappear between 'squeezes'.

What's happening?

Molecules of water vapor will bounce around like molecules of other gases unless you give them a reason to stick together. Cooling the vapor slows the molecules down, so they have less kinetic energy and more time to interact with each other. How do you cool the vapor? When you squeeze the bottle, you compress the gas and increase its temperature. Releasing the container lets the gas expand, which causes its temperature to go down. Real clouds form as warm air rises. As air gets higher, its pressure is reduced. The air expands, which causes it to cool. As it cools below the dew point, water vapor forms the droplets we see as clouds. Smoke acts the same in the atmosphere as it does in the bottle. Other nucleation particles include dust, pollution, dirt, and even bacteria.

148)   BUBBLES AND TEMPERATURE

The purpose of this project is to determine if temperature affects how long bubbles last before they pop.

Materials you'll need:

- Clear jars with lids
- ½ cup dishwashing liquid
- 4 ½ cups water
- 4 Tbsp. glycerin
- measuring spoons
- thermometer
- stopwatch or clock with a seconds hand

What to do:

1. First make a bubble solution, by combining and gently mixing the dishwashing soap, water, and glycerin.
2. Use your thermometer to find places that are different temperatures from each other.
3. Label each jar with either where you are placing it, or the temperature.
4. Add the same amount of bubble solution to each jar and replace the lid
5. Place the jars at the locations with different temperatures and wait at least 30 minutes.
6. After 30 minutes, shake each jar for 30 seconds and record how long it takes for all of the bubbles to pop.
7. Do one jar at a time and record the temperature and the total time it took for the bubbles to pop.

What's happening?

The *relative humidity* inside closed jars is higher at warmer temperatures, which affects how long bubbles last after you shake them.

149)   CREATE A TORNADO (METHOD 1)

In this fun experiment you will learn how to make a mini tornado using everyday items you can find around the house.

Materials you'll need:

- Water
- A clear plastic bottle with a cap
- Dish washing soap
- Optional: glitter

What to do:

1. Fill the plastic bottle with water until it is ¾ full.
2. Add 3-4 drops of dish washing liquid.
3. If you have glitter, sprinkle in a few pinches.
4. Put the cap on the bottle and make sure it is on as tightly as can be.
5. Turn the bottle upside down and hold it by the neck.
6. Quickly spin the bottle in a circular motion for a few seconds, stop and look inside! (If it doesn't work the first time, don't worry! Just keep trying).

What's happening?

As you spin the bottle, the water inside up against the plastic is pulled along due to its friction again the plastic walls. The water towards the inside of the bottle takes longer to get moving. However, if you keep spinning, both the bottle and the water are spinning together. When you stop moving the bottle, the water inside keeps spinning (this is called *centripetal force).* Your mini tornado can be seen for just a few seconds when the outer water slows down and the inner water continue to spin rapidly. The circular path on the inside is called the *vortex.*

150)   CREATE A TORNADO (METHOD 2)

Create another homemade tornado, this time using different materials, and creating a different reaction!

Materials you'll need:
- Tall drinking glass
- Bottle of soda water (carbonated water)
- Table salt
- Long spoon or stirrer

What to do:

1. Pour the soda into the glass until it is almost full (about 2/3 of the way)
2. Use the spoon to stir the liquid as fast as it can go (without spilling).
3. As the water moves pour some salt into it.
4. Watch as tornado forms!

What's happening?

Tornadoes happen when warm, moist air mixes with cool, dry air. When there are clouds with spirals of rising air (aka *vortex)* inside. The vortex sucks in cool, dry air, and more air moves in around it to fill the space. When you stir the carbonated water or soda, you are creating a *vortex.* By adding salt, you help the *carbon dioxide (CO2)* escape from the soda water, which allows for more and more bubbles to form. The faster the bubbles form, the faster the cool, dry gas of CO_2 can escape and help create a tornado.

151)   HOMEMADE LIGHTNING

This is another fun science experiment that teaches you about static electricity (and weather science!).

Materials you'll need:

- pencil (with an eraser)
- aluminum tray (a pie tin works great)
- wool cloth
- Styrofoam plate
- Thumbtack

What to do:

1. Stick the thumbtack through the bottom of the aluminum tray then stick the pencil eraser to the thumbtack on the other side. This will make the pencil a handle for the tray.
2. Rub the wool on the Styrofoam plate hard and fast for AT LEAST two minutes.
3. Pick up the aluminum tray using the pencil handle and set it upside down on the Styrofoam plate.
4. Watch for sparks!

What's happening?

By rubbing the wool on the aluminum tray, *electrons* (negatively charged particles) come off the wool and onto the plate. Styrofoam is an *insulator*, which means it is a material that doesn't let electrons pass through it. However, aluminum tray is a *conductor,* which means it is a material that does allow electrons to move through it. The electrons leave the Styrofoam plate and jump to the aluminum tray, making big sparks!

152)   CREATE WIND

This wind experiment uses a hot oven so be sure to ask for adult help. Have fun and be safe!

Materials you'll need:

- Two metal baking pans
- Two heatproof pads or boards (wooden boards work great)
- Oven mitt
- Large cardboard box
- Sand
- Ice
- Incense stick
- Scissors
- Matches or a lighter

What to do:

1. Fill the first pan with sand and place in an oven on the lowest heat setting. Ask an adult for help with the oven.
2. Use the scissors to cut off the front of the cardboard box.
3. Place the heatproof pads or boards next to each other inside the box.
4. Fill the second pan with ice and place it on one of the pads.

5. Wait 10-15 minutes for the first pan to warm sufficiently, then use the oven mitt to remove it from the oven (ask for help!).
6. Place it on the pad next to the pan of ice so that the two pans are practically touching.
7. Light an incense stick and hold it horizontally, with the burning tip right between the two pans.
8. Where does the smoke go?

What's happening?

Smoke from the stick will always drift toward the warm pan full of sand. The smoke is riding the "breeze" that you have made with the two pans. The pan full of ice cools the air above it, which causes air to sink. The pan full of warm sand warms the air above it, which causes to rise. This creates a *low-pressure* zone just above the sand, and the cool air above the ice will rush in to equalize the pressure. This is how wind works! For example, on hot days, land (your tray of sand) will heat up faster than water (your tray of ice), so the cooler air above lakes and seas will move towards land, which creates a breeze.

153)   MEASURE WIND SPEED

Learn how to make your own anemometer, and measure the wind speeds around your house!

Materials you'll need:

- Paper plate
- Four paper cups
- Colored tape
- Pushpin with a plastic end
- Pencil with an eraser
- Double-sided tape
- Stopwatch or clock

What to do:

1. Mark one paper cup by using the colored tape to wrap around the outside.
2. Attach a short piece of double-sided tape to the side of each cup. Attach one cup to the edge of the plate with the opening facing left.
3. Attach the rest of the cups the same way all around the plate.
4. Put the pushpin through the very middle of the plate, pointing down.
5. "Pin" the plate to eraser on the pencil (this will be your handle).
6. Now lift up the plate, using the pencil handle, and he plate and cups so they may spin in the breeze.

7. Look up the wind speed in your area.
8. Count how many times the cup with the colored tape on it spins around in 30 seconds. Write it down.
9. Repeat the process at different times over the next week, and soon you'll be able to

What's happening?

A professional anemometer uses spinning discs linked to an electronic device that counts the turns. You have made your own! The wind catches the cups and turns them, the number of *revolutions* (turns) per minute gives you the speed of the wind.

154)   LEARN ABOUT THE GREENHOUSE EFFECT

Materials you'll need:

- Large plastic bottle with lid
- Glass jar small enough to fit inside the bottle
- Scissors
- Thermometer

What to do:

1. Remove the label from your plastic bottle but keep the lid screwed on.
2. Use your scissors to cut off the bottom part of the plastic bottle.
3. Place the thermometer inside the jar, and set it in a sunny spot.
4. Wait and hour and then check the temperature. Record the number.
5. Now place the plastic bottle over the jar and leave it for another hour.
6. Check the temperature again, record it, and compare it to the earlier temperature.

What's happening?

You'll notice that the second temperature reading is much warmer than the first. That's because the energy from the sun (solar energy) that is passing into the plastic bottle has been turned into heat that can't escape. This process is called the greenhouse effect. The atmosphere of the Earth acts like the plastic bottle—it allows the sun's energy to pass through and warm the planet, but then heat is then trapped within the atmosphere.

155) CONVECTION CURRENTS OF COLOR

Do you know what convection currents are? Don't worry, this colorful experiment will teach you all you need to know!

Materials you'll need:

- Four identical empty plastic bottle (with openings of at least 1 ½" in diameter).
- Warm and cold water
- A sink
- Yellow and blue food coloring
- 3 x 5 index card (or a playing card)

What to do:

1. Fill two bottles with warm water from the tap and the other two bottles with cold water.
2. Add yellow food coloring to the warm water bottles.
3. Add blue food coloring to the cold water bottles.
4. Make sure each bottle must be filled to the very top with water.
5. Now, place the index card over the mouth of one of the warm water bottles.
6. You may want to do this over a sink—hold the card in place as you turn the bottle upside down and place it on top of one of the cold water bottles (so they are mouth-to-mouth).
7. While holding the top bottle in place, CAREFULLY slip the card out from in between the two bottles.

8. Observe what happens.
9. Now, repeat steps 5-7, but this time place the bottle of cold water on top of the warm water.
10. Again, observe what happens.

What's happening?

Warm water is less dense (lighter in weight) than cold water. When you place the bottle of warm water on top of the cold water, the denser cold water stays in the bottom bottle and the less dense warm water stays in the top bottle. However, when you switch the two, and place the cold water bottle atop of the warm water, the less dense warm water rises to the top bottle and the cold water sinks. You can clearly see the movement of water as they mix and create green water. The movement of warm and cold water inside the bottles called a *convection current*. Convection currents in our oceans and atmosphere create thunderstorms as the warm and cold air masses collide and mix.

156) PRESSURE PAPER

Ask an adult for help finding the pieces of wood paneling.

Materials you'll need:

- Several pieces of wood paneling (1" x 36" x ¼" thick)
- Several sheets of newspaper
- Work gloves
- Table

What to do:

1. Place the piece of wood on a table and let one end hang over the edge by 4 inches. Now, while wearing work gloves, karate-chop the edge of the wood hanging off the table. What happens? (Make sure people are out of the way).
2. Place the stick on the table again, this time sticking 4-10 inches hang over the edge.
3. Fold a single sheet of newspaper 3-4 times. Place the folded newspaper over the end of the stick that is lying on the table. Now karate chop the edge of the stick again. What happened? (Again, make sure everyone is standing away from the table).
4. Replace the stick how it was in Step 2.
5. Finally, take a new sheet of newspaper and use it to cover the portion of the stick that is lying on the table. DO NOT FOLD IT, and make sure that the newspaper is flush with the edge of the table.

6. Smooth down the newspaper with your hands so that
    there are no pockets of air under the sheet of paper
    (important!!)
7. Wearing your gloves, strike the protruding edge of the
    stick with a sudden sharp hit.
8. What happens?

What's happening?

Though the weight of the folded paper and the flat paper
are the same, there are other forces at work—namely *air
pressure.* It is the pressure of the air pushing downward on
the newspaper that prevents the paper from rising. The flat
paper has more surface area for the air pressure to push
down upon. Smoothing down the newspaper with your
hands, ensures that there is no air under the newspaper that
may help it lift up when you hit the stick.

XIV. PLANTS AND FOOD SCIENCE

157)   HOW TO GROW BACTERIA

Bacteria are everywhere—it grows in dirt, water, plants animals, and even humans! Thankfully, our immune system works to make bacteria harmless. Bacteria are a fascinating type of microorganism, and this experiment shows you how to grow your own sample while also learning how it behaves.

Materials you'll need:

- Petri dish of agar
- Cotton swabs
- Some old newspaper

What to do:

1. Using your cotton swab, collect a sample from somewhere in your house. Simply rub the cotton swab on any surface you like.
2. Take it to you Petri dish with agar, and rub the swab over the agar with a few gentle.
3. Replace the lid and seal the Petri dish.
4. Place the Petri dish in a warm area for 2 to 3 days.
5. Check the dish every day to see the growth of the bacteria. Track it by drawing what it looks like each day, and write down the changes.

6. To properly dispose of the bacteria, wrap up the Petri dish in old newspaper (don't open the lid) and placing in a trash can.

What's happening?

The Petri dish of agar and warm conditions provide the ideal environment for bacteria to grow. The microorganisms in the dish grow into individual *colonies*, each a clone of the original. The original bacteria you collected grows steadily, becoming visible to the naked eye in 2-3 days.

MORE: Repeat the process with a new Petri dish, and take a sample from under your finger nails or between your toes. How is this sample different than the first? You can repeat this experiment as many times as you like, just be sure to use a new Petri dish each time.

158)   DYED FLOWERS

In this experiment find out how plants "drink" water, a process called capillary action. You can use carnations, roses, or even celery too see it work!

Materials you'll need:

- Three white flowers—anything from your florist will do: roses, carnation, daisies, even celery!
- Food coloring—4 different colors.
- 4 drinking glasses
- Water
- Scissors or a sharp knife

What to do:

1. Add many drops of food coloring to a glass of water. (Make a couple different glasses with different colors).
2. Add one flower to each of the glasses.
3. Take the last flower and slit the bottom part of the stem in two.
4. Now take the two remaining glasses of dyed water place each half of the stem into different colored glasses.
5. Keep the flowers and glasses away from sunlight and wait.

What's happening?

This is a lesson in *transpiration*, which means that the plant draws water up through its stem. When the water is drawn up through the stem it goes to leaves and petals, which is why the petals turn a different color. Neat, huh? You've made your own type of flower!

159)   GERMINATE SEEDS AT HOME

Plant some seeds and follow their growth as they sprout (a process called *seed germination)*, while learning what plants need in order to grow.

Materials you'll need:

- Fresh seeds (Whatever kind you want! Sunflower seeds, bean seeds, radish seeds, etc.)
- Good quality soil (Can be found at any local garden store).
- A container for the soil and your seeds.
- Water, light, and heat.

What to do:

1. Fill the container with soil.
2. Plant the seeds in the soil.
3. Place the container somewhere warm (a window sill a great spot).
4. Keep the soil moist by watering it whenever it feels dry. (Be careful not to overwater).
5. Wait and watch as the seeds germinate and seedlings begin to sprout over the next week or so.

What's happening?

Germination is the process of a plant sprouting from a seed and beginning to grow. In order for seedlings to germinate, they need proper conditions. Most need a warmer temperature (which is why you put your container on the window sill). Water and oxygen are also needed for seed germination. These are the basics, but some seeds are different, so read the instructions on the back of each seed packet to see their specific requirements.

MORE: Continue to look after your seedlings and monitor their growth in the following weeks. Write down and compare the growth rates of your different types of seeds.

160)   SEED BALLS

Making seed balls will help your seeds grow in areas that are barren or unfertile.

Materials you'll need:

- Modeling clay
- Plant seeds (whatever kind you want!)
- Potting soil
- Parchment paper

What to do:

1. Knead the clay and then flatten it and shape it into a disc shape.
2. On top of the clay, add ½ teaspoon of potting soil and ¼ teaspoon of plant seeds.
3. Fold the clay inward, keeping the soil and seeds from spilling out.
4. Mold the clay into a ball around the soil and seeds.
5. Mix some more soil and seeds together on a flat surface.
6. Roll your clay ball through the soil and seeds, covering it evenly.
7. Knead the soil and seeds into the clay so that they become fully incorporated into the clay ball.
8. Repeat until you have as many as you want.
9. Place them on some parchment paper for a few days to dry.

10. Once dry, simply take them outside and toss them
    where you want plants to grow.

What's happening?

Seeds need warmth, protection, and nutrients to grow. The
clay vessels that you make for your seeds and soil offer
provides all these things. The seeds remain safely inside,
until rains soak the clay and stimulates the seeds' growth
and germination.

161)   FOOD SAFETY EXPERIMENT

This experiment will help remind you to wash your hands before touching food! Get ready to get grossed out!

Materials you'll need:

- 3 slices of bread (any kind will do)
- 3 sandwich bags (that can be sealed)
- Sharpie pen (or other marker)

What to do:

1. Label one bag "control," the second "clean hands" and the third "dirty hands."
2. Without washing your hands, touch the first slice of bread. Feel free to touch it all over, and even pass it along to your family and friends to touch.
3. Place it in the "dirty hands" bag and seal it.
4. Now wash your hands thoroughly and handle the second piece of bread.
5. Place this in the "clean hands" bag and seal it.
6. Without handling it too much, place the last piece of bread the "control" bag and seal it
7. Put all 3 bags in a cool dry place and record daily observations of mold growth. You may be surprise at how quickly it shows up!

What's happening?

Mold will grow on all the slices, but more quickly and with more on the "dirty hands" sample. The germs on dirty hands became food for the mold to grow. Remember to wash your hands before you handle food!

162)   MUMMIFY A CHICKEN!

This experiment will show you how Ancient Egyptians made mummies. It involves raw meat, so be sure to get help from an adult!

Materials you'll need:

- Cornish game hen (or other smaller bird)
- Salt
- baking soda
- gallon-size plastic bag
- lots of paper towels
- large bowl

What to do:

1. Wash the chicken
2. Thoroughly dry it with the paper towels
3. In the large bowl, mix equal parts baking soda and salt.
4. Put the chicken in the large plastic bag and pour in the salt and baking soda mixture until it is completely covered.
5. Place the chicken out direct sunlight and let it sit in the baking soda and salt mixture for a week or more.
6. After a week, pour the salt and baking soda mixture out of the bag and check out your mummified chicken!

What's happening?

Salt and baking soda are considered *desiccants,* which mean they dry things out. In order for a body (or chicken) to decay, it needs the liquids in the body to cause decomposition. The salt and baking soda mixture draw out the moisture and prevent the chicken from decomposing.

XV. MAKE YOUR OWN....

163)   MAKE YOUR OWN MOVIE (WITHOUT A VIDEO CAMERA)!

Make a homemade *thaumatrope*! What is a thaumatrope you ask? Follow the directions to find out!

Materials you'll need:

- small piece of cardboard (2in. x 3in)
- sharpened pencil or pen
- two pieces of string
- tape
- two pieces of paper

What to do:

1. Take your piece of cardboard and, using your sharpened pencil or pen, punch two holes in the cardboard. One on the left side, and one on the right.
2. Write a small "x" in the center of the cardboard.
3. Cut out two pieces of white paper that are a little smaller than the cardboard piece. (Be sure the holes still show).
4. On one piece of paper, draw a fish. On the other, draw a larger fishbowl.
5. Tape the picture of the fish on one side of the cardboard, so that the fish is directly over the "x".

6. Tape the fishbowl on the other side, making sure the bowl is also centered over the "x".
7. Thread one piece of string through the hole on one side. Thread the other piece of string through the hole on the other side.
8. Now twist the strings together on each side.
9. When the strings are all twisted up, give them a pull (as if you are trying to stretch a rubber band). This will make the cardboard twirl.
10. Watch the show, what do you see?

What's happening?

You have made a thaumatrope, which was a toy very popular in the 19th century. It seems like the two images blend, because your eye continues to see the image of the fish after it is gone. This is called *persistence of vision.* During this time, the picture of the fishbowl is twirled quickly into sight, and you seem to see both pictures at once, overlapping them in your mind. This is the same principle with movies—movie film is simply a series of pictures separated by black spaces, but they are flashed on the screen so quickly that your vision processes them as a smooth, continuous picture.

164)   HOW TO MAKE YOUR OWN BATH SALTS

Whether you're making a special present for someone else, experimenting at home or just want to relax in a hot bath, give this experiment a go. Create your own bath salts with a variety of refreshing fragrances, experiment with different essential oils to see which you like best.

Materials you'll need:

- 1 cup of washing soda
- A plastic bag
- A rolling pin
- A bowl
- A spoon for stirring
- Essential oil
- Food coloring

What to do:

1. Take the cup of washing soda and put it into a plastic bag.
2. Crush the lumps with a rolling pin.
3. Empty the bag into a bowl and stir in 5 or 6 drops of your favorite essential oil.
4. Stir in a few drops of food coloring until the mixture is evenly colored.
5. Put the mixture into clean dry containers and enjoy as you please.

What's happening?

Bath Salts are typically made from Epsom salts (*magnesium sulfate*) and washing soda (*sodium carbonate*). The chemical make-up of the mixture makes it easy to form a lather. Bath salts are said to improve cleaning and deliver an appealing fragrance when bathing.

165) GROW YOUR OWN CRYSTALS

Growing crystals is a simple process and involves a lot of scientific principles. This project will require the help of an adult.

Materials you'll need:

- ¾ cup of Alum powder
- Egg shells (real or plastic)
- Glue
- Paintbrush
- water
- Box
- Large bowl
- Spoon
- Optional: food coloring

What to do:

1. One day before, paint the eggshells with a *thin* layer of glue. If you want crystals to grow around the edge, be sure to paint a little glue on the outside of the rim of the eggshell.
2. While the glue is still wet, sprinkle the egg with alum powder, let it dry overnight.
3. If you are using dye, stir it into 2 cups of water.
4. With the help of an adult, heat the water until it simmers.
5. Pour the alum powder into the water and stir until it is completely dissolved. The idea here is that you want a saturated solution. In other words, you want the water

to be holding as much alum powder as it can possibly hold, but you don't want a lot sitting on the bottom.
6. When the mixture is sufficiently saturated, drop an eggshell into the mixture.
7. Leave it overnight (or longer for larger crystals).

What's happening?

All substances have a preferred crystal shape, depending on the alignment of the atoms or molecules that make it up. Potassium aluminum sulfate's crystal shape is isometric. In other words, this alum likes to make octahedrons. If these crystals are crowded, they grow into whatever space is available, so you will see a lot of different shapes growing!

166)   MAKE YOUR OWN SNOWFLAKE

Find out how crystals are formed in this fun science experiment, and you can even keep your finished snowflake as a cool decoration. This project uses hot water, so be sure to ask an adult for help.

Materials you'll need:

- String
- Scissors
- Wide-mouth jar or large glass
- White pipe cleaners
- Boiling water (ask for help!)
- Borax
- Small wooden stick (or pencil)
- Optional: blue food coloring (or whatever color you want!)

What to do:

1. Take one pipe cleaner and cut it into three sections equal-sized pieces.
2. Twist these sections together in the center so that you now have a shape that looks something like a six-sided star. (If the points of your shape are uneven, trim them to match).
3. Take the top of one of the pipe cleaners and attach a piece of string to it.
4. Tie the other end of the string to your wooden stick.

5. With the help of an adult, carefully fill the jar with boiling water
6. For each cup of water you use, add 3 Tablespoons of borax (adding one tablespoon at a time).
7. Stir well until the mixture is mostly dissolved (it's okay if some of the borax settles at the bottom).
8. If you have food coloring, add a few drops now.
9. Place your snowflake in the jar so that the wooden stick is resting on the top of the jar and the snowflake is sitting in the liquid solution.
10. Leave the whole thing soaking overnight, and check it in the morning. What does it look like?

What's happening?

Borax (aka sodium borate) is a white powder made up of colorless crystals that dissolve easily in water. When you add borax to the hot water you can dissolve more than you could if you were adding it to cold water (remember the **Hot Sugar Experiment?).** As we learned in that experiment, hot water molecules move around fast have a lot of space between them which allows more borax crystals to dissolve.

When you let the solution cool overnight, the water molecules move closer together and can't hold as much of the borax. Crystals begin to form on top of each other and on top of your pipe-cleaner shape, and soon it looks like a completed crystal snowflake! Try all kinds of colors of food coloring for different kinds of snowflakes, and use the string to hang it up as a cool science decoration.

167)   MAKE YOUR OWN INVISIBLE INK (METHOD 1)

Making invisible ink is easy and a lot of fun. You can use it to write secret messages that no one else can read!

Materials you'll need:

- Half a lemon
- Water
- Spoon
- Bowl
- Cotton swab
- White paper
- Lamp

What to do:

1. Squeeze lemon juice into the bowl.
2. Add a few drops of water and mix well.
3. Dip the cotton swab into the mixture and write your message onto the white paper.
4. Wait for the juice to dry (once it's dry it will be totally invisible).
5. When you want to reveal your message, hold the paper close to a light bulb.

What's happening?

Diluting the lemon juice with water makes it very difficult to see when it's on paper and no one can see it. However, lemon juice *oxidizes* and turns brown when heated. By putting it near a light bulb, it heats up, turns brown, and then you or your friend can read the secret message.

168) MAKE YOUR OWN INVISIBLE INK (METHOD 2)

Here's another way to make invisible ink, this time using a mirror, a soapy solution, and a foggy room.

Materials you'll need:

- liquid dishwashing detergent
- 1 cup of water
- a few cotton swabs
- bathroom mirror (or hand mirror)

What to do:

1. Place a 2-3 drops of dishwashing soap into your cup of water. Stir well.
2. Dip a cotton swab into the soapy solution and write a note on the mirror.
3. When the solution dries it should be invisible.
4. When you take a hot shower or bath, close the door so the foggy steam stays in the bathroom (or near the handheld mirror)
5. Watch what happens. Can you see your message?

What's happening?
When you take a bath, the foggy steam that forms on the mirror is made up of water molecules. These tiny drops of water stick together because of a force called *surface tension*. However, your invisible ink (the soapy solution) breaks the surface tension of the water. So wherever you

used your soapy "ink", the water molecules are unable to form into droplets on the mirror.

MORE: Write a message on the bathroom mirror and wait until someone else takes a shower or bath. Wait and see how long it takes them to discover your note!

169)   MAKE YOUR OWN SUGAR CHRISTMAS ORNAMENTS

This experiment uses a microwave and hot substances, so be sure to ask for help from an adult before you start!

Materials you'll need:

- 3 cups of granulated sugar
- 1 cup of water
- Pipe cleaners
- Pencil
- Food coloring
- Glass jar
- Parchment paper
- Microwave
- Microwave safe container

What to do:

1. Pour the sugar and the water into a microwave safe container and stir.
2. Microwave your sugar solution on high for 2 minutes. With the help of an adult, CAREFULLY Use remove the solution from the microwave and stir it again.
3. Microwave the solution 2 minutes more, carefully remove it, and stir.
4. Add several drops of food coloring and stir it into the solution.
5. Pour the solution into glass jar and allow it to cool to room temperature.

6. Create an ornament shape using pipe cleaners (like a candy cane, snowflake, or stocking).
7. Wrap the end of the pipe cleaner around the center of a pencil.
8. Hold onto the pencil and dip your ornament into the sugar solution.
9. Lay your ornament flat on a sheet of parchment paper to dry.
10. Once it has dried, place the ornament back into the solution. Place the pencil on the mouth of the jar, so that the ornament hangs down without touching the sides or bottom of the jar.
11. Let it sit like that for at least one week.
12. Pull the ornament out of the jar, let it dry on the parchment paper.

What's happening?

When you put 3 cups of sugar into 1 cup of water, you create a *supersaturated solution*. This means there is more *solute* (the sugar) than the *solvent* (the water) can hold. By heating up the solution you are able to dissolve as much sugar into the water as possible (due to the microscopic expansion of the water molecules).

Dipping your pipe-cleaner ornament into the solution and allowing it to dry creates a small layer of sugar crystals around the pipe cleaner. Though these sugar crystals were "dissolved" in the solution, the pipe cleaner gives them a surface they can use to *recrystallize*. Over the course of a week, you can see just how much sugar from your *supersaturated solution* is crystallized on your ornament

170)   MAKE YOUR OWN SILVER POLISH

Do your parents have any tarnished silverware or silver pieces in the house? "Wow" them by making your own solution to make their dirty silver shine again (but always ask before you do so)! This experiment does use hot water, so be sure to get help from an adult.

Materials you'll need:

- a tarnished piece of silverware or any other kind of silver
- a large pan
- aluminum foil
- ½ gallon of hot water (ask for help!)
- ½ cup of baking soda
- sink

What to do:

1. Line the bottom of the pan with foil.
2. Place the silver object on top of the aluminum foil (be sure the silver touches the aluminum).
3. Take your ½ gallon of hot water and add ½ cup of baking soda. (Do this in the sink, as the mixture will froth and might spill over).
4. Pour the hot baking soda and water mixture into the pan until the silver is completely covered.
5. The tarnish should start to disappear! (If the silver is super tarnished, you may have to repeat this process a few times).

What's happening?
When silver tarnishes, it combines with sulfur and forms *silver sulfide,* which gives it the black color and darkens the silver. However, the silver can be returned to its shiny state by removing the silver sulfide coating from the surface.

Combining the *silver sulfide,* aluminum, and hot baking soda solution creates a chemical reaction. The baking soda solution takes the sulfur from the silver and carries it to the aluminum. The *aluminum sulfide* may stick to the foil, or it may make little yellow flakes in the bottom of the pan. The silver and aluminum must be touching each other, because a small electric current flows between them during the reaction, which is called an *electrochemical reaction.*

171) MAKE YOUR OWN BALLOON POWERED CAR

In this experiment, you use the power of air pressure to power your car and learn about Newton's Third Law of Motion. This experiment uses scissors and wire cutters, so be sure to ask an adult for help!

Materials:

- Foam core or corrugated cardboard
- Wooden barbeque skewers
- Regular cardboard
- Plastic straws
- Tape
- Balloons
- Scissors
- Wire cutters

What to do:

1. Begin by cutting the "chassis" of your car. Cut a 6×3 inch piece of your corrugated cardboard (or other material) using the scissors.
2. Next, the axles. Use wire cutters to snip two 4" pieces of your wooden barbeque skewer.
3. Now you something to mount the axles on the chassis. For mounts, cut two 3"sections of straw and use tape to fasten the mounts to the front and back (the smaller sides) of your chassis.

4. To mount your axles, simply slide the wooden skewers
   through the middle of the straws.
5. Now use scissors to cut four pieces of regular cardboard
   into quarter-sized pieces. These are the wheels! (To
   make it easier, trace a quarter onto a piece of cardboard
   and then cut out the shape!)
6. To mount your wheels, push the circles onto the
   skewers, one on each end of both skewers.
7. Now for the balloon engine—cut the lip off the balloon.
8. To make your "exhaust pipe," insert straw 1" into the
   balloon. Use tape to fasten the straw inside the balloon,
   and make it tight! The tighter the seal, the better your
   experiment is going to work.
9. Attach the exhaust pipe in such a way that the point
   where the straw and balloon connect is 1" from the end
   of your chassis and the straw is pointing straight out.
   Tape it securely.
10. Blow up the balloon through the straw and pinch the
    straw to keep the air from escaping.
11. Put your car on the ground and let go!

What's happening?

This experiment demonstrates *Newton's Third Law of
Motion*, which states that for every action, there is an equal
and opposite reaction. The escaping air from the balloon
rushes out of the straw which causes *propulsion*. This is the
action, and the *reaction* is the car moving! The *potential
energy* (stored energy) of the car is in the balloon filled
with air. As the air flows from the balloon and moves the
car, it changes to *kinetic energy (*moving energy).

172)   MAKE YOUR OWN POWER DOOR OPENER

How cool would it be to have a power door opener as seen in science-fiction and spy movies? This project will show you how to use a toy car to do the trick. A small wire-controlled car has enough power to push and pull a typical room door back and forth if you know the super-sneaky way to install it.

Materials you'll need:
- Old remote or wire-controlled toy car
- Velcro fastening tape, adhesive backed
- Screwdriver
- Pliers

What to do:

1. Remove the body shell from the toy car with a screwdriver.
2. Remove the front wheel and axle.
3. Use the fastening tape, attach the car near the bottom end of the door.
4. Using the remote control, see if it can push the door open or closed. If not, try to reposition the car.

What's happening?

Once you find the proper position, you will be able to open the door with your car! Your remote controlled car is working how it always does, but since it is against the door, it uses the traction of the floor to push it open or closed!

173)   MAKE YOUR OWN VIDEO PROJECTOR

Use the science of magnification (and the help of an adult) to make your own movie projector.

Materials you'll need:

- Cardboard box
- Smartphone
- Packing tape
- Magnifying glass
- Utility knife
- Scissors

What to do:

1. On one of the shorter ends of the box, trace a circle around the magnifying glass.
2. With adult help, use the utility knife to cut out the traced circle.
3. Cut off the short flap on the opposite side of the hole.
4. Fold the flap into a prop for a small phone.
5. Place the stand inside the box.
6. Tape the magnifying glass over the hole you had cut.
7. Set your screen to a landscape orientation, pick a video, and start playing it.
8. Place your phone upside down on the "stand" and seal the box with packing tape.
9. Turn off all the lights and watch the show!

What's happening?

We'll start with the upside-down phone. The human eye has a lens (like the magnifying glass), and the lens flips the image right-side up through *refracting* the light from the phone's screen, much like your eye does the light from the world. While our brain flips the image back to its right-side-up orientation, the wall doesn't know how to do that, which is why you put the phone in the box upside down. Even though the magnifying glass is smaller than the screen it captures the whole image because the shape of the lens is *convex*. This means the sides bend outwards, rather than inwards (*concave*). The shape of this lens allows it to catch and focus more space from inside of the box. When you make the room dark, the magnifying glass with gather and focus all of the light from the box and you can see it all reflected on the wall.

174)   MAKE YOUR OWN COMPASS

This experiment uses sharp objects so be sure to ask an adult for help.

Materials you'll need:

- plastic lid (or shallow bowl)
- needle
- a bar magnet
- A cork (or Styrofoam)
- knife
- water

What to do:

1. With the help of an adult, take the "North" side of your bar magnet and place it perpendicular to your needle.
2. Hold the needle in one hand and slide one side of the bar magnet along the length of the needle. Don't slide the magnet back and forth on the needle; instead, just slide it in one direction, from one side to the other every time. Do at least 50-75 times "activate" your needle magnet.
3. Turn the needle upside down, and take the opposite end of the magnet and repeat the process.
4. With the help of an adult, slice the end of the cork off, so it has the thickness of a quarter.
5. Pour some water into your shallow dish.
6. Place the needle on top of the water and watch as the tip of the needle spins to point North!

What's happening?

By rubbing the different ends of needle with the bar
magnet, a little bit of its magnetic material will be left
behind on the needle, which makes the needle a magnet! In
order for the needle to act as a compass, it needs a place
with less *friction* so that it can move. By making a little
raft in the bowl of water with the cork material (which is
less dense than water, hence the floating), the needle can
freely move to point North.

175)   MAKE YOUR OWN HAND WARMERS

Have you ever used or seen chemical hand warmers? With
this science experiment, you can make your own while
witnessing how a chemical reaction like rusting can be used
for keeping your hands warm. This experiment will create
a heat chemical reaction, so be sure to get an adult's help
with this one.

Materials you'll need:

- Jelly Crystals
- Small cups
- Iron filings
- Calcium chloride
- Ziploc bag
- Water

What to do:

1. Fill a cup with 9 ounces of water.
2. Add a big scoop of Jelly Crystals to the water and
   wait until they are fully grown. (Hint: they're done
   growing when the water in the cup is gone).
3. Add 4 Tablespoons of the full-grown Jelly Crystals to
   a plastic Ziploc bag.
4. Add 1 tablespoon of iron filings to the bag.
5. Add 1 ½ Tablespoons of calcium chloride (ice melt
   pellets).
6. Carefully mix the contents of the Ziploc bag by
   squishing the bag with your fingers.

7.   Once everything is thoroughly mixed, seal up the bag.
8.   Shake and squish the bag to mix the contents even more.
9.   What do you notice about the bag?

What's happening?

You should feel the heat coming from the bag. The chemical reaction between the iron, calcium chloride salt, air, and water (which was locked inside the Jelly Crystals) produces *iron oxide* and heat. The heat is a result of the iron oxide chemical reaction. Because the chemical reaction creates heat, it is called an *exothermic reaction.*

176)   MAKE YOUR OWN SOLAR OVEN

Make s'mores with a solar oven!

Materials you'll need:

- Cardboard box
- aluminum foil
- tape
- plastic wrap
- Graham crackers
- Marshmallow
- Chocolate bar
- A bright, sunny day.

What to do:

1. Line the inside of a cardboard box with foil, the bottom and the sides.
2. Place the chocolate bar and marshmallow on top of the graham cracker. And place inside the cardboard box, on top of the foil
3. Cover the box with plastic wrap and secure it with tape.
4. Carefully move the box into the sun for 1-2 hours.
5. Enjoy your treat!

What's happening?

The foil that is lining the inside of the box reflects the sun's rays (and heat) very strongly. The sun's light and heat warms the s'mores and the aluminum foil dramatically increases the heat and light, "cooking" the s'mores.

177)   MAKE YOUR OWN ICE CREAM!

Ask for help from adult with this extremely delicious experiment.

Materials you'll need:

- 2 large Ziploc freezer bags
- Measuring utensils
- Large container with a lid
- Crushed ice
- Rock salt
- gloves
- 1 teaspoon Vanilla extract
- ½ cup Half and half
- 1 Tablespoon brown sugar

What to do:

1. Fill the large container about half full with crushed ice and add about 6 Tablespoons of rock salt. Put the lid on it, put on your gloves, and shake the ice and salt for about five minutes.
2. Add the half and half, brown sugar, and vanilla extract to a large freezer bag.
3. Seal the bag tightly, allowing as little air to remain in the bag as possible. Place the sealed bag inside the other bag and seal it well (this will minimize leaking).
4. Place the two bags inside the container with the ice and seal the container.

5. Mix that plastic container like crazy! You'll want to do this for about 15-20 minutes, so it will be nice to have someone else to help.
6. Once mixed, remove the bags from the jar and rinse them with water to prevent any salt water from getting into your ice cream.
7. Pour into a bowl and enjoy!

What's happening?

By adding salt to the ice, you lower the freezing point of the ice. When salt is added to the ice, some of the ice melts because the freezing point is lowered. Heat must be absorbed by the ice for it to melt, and in this experiment, the heat comes from the cream mixture. By lowering the temperature at which ice is frozen, you create an environment where the cream mixture can freeze at a temperature below 32° F and turn into delicious ice cream.

XVI. MISCELLANEOUS EXPERIMENTS (BALLOONS, COLORS, MORE!)

178)   THE "UN-POPPABLE" BALLOON

In this experiment, you'll learn how to stick a needle into a balloon without popping it! Wow your friends with your "magical" needle and balloon trick!

Materials you'll need:

- Some balloons
- long wooden or metal skewers
- petroleum jelly

What to do:

1. Blow up a balloon (not too full) and tie the opening shut.
2. Dip the tip of one of your skewers in petroleum jelly and spread it along the entire length of the skewer.
3. Now, CAREFULLY insert the skewer with a gentle twisting motion into the top of the balloon (opposite the knot).
4. Slowly push and twist the skewer all the way through the balloon, until the tip emerges from the other end, near the knot. It shouldn't pop!
5. Now, repeat the process with another balloon. Try to stick the skewer into the side of the balloon.

6.   What happens?

What's happening?

The rubber in the balloon consists of many long molecules that are linked together, called polymers. When molecules of a polymer are chemically attached to each other (like in the rubber of a balloon) it is called *cross-linking*. These links hold the polymer molecules together and allow them to stretch. The rubber at the ends of the balloon is stretched out less than in the middle of the balloon. This means there is less force pulling on those parts. This allows the tip of the skewer to break SOME of the polymer cross-links, push aside the molecules of rubber, and slide into the balloon. Enough cross-links remain so that the balloon holds together.

However, on the side of the balloon, there are fewer polymer molecules, so when you push the tip of the skewer through the side, the skewer breaks a few of the cross-links, the tension on the remaining links is too much, and POP!

179) SPINNING PENNY

Materials you'll need:

- Clear, latex balloon
- Penny
- Air pump

What to do:

1. Drop the penny into the balloon and make sure it goes all the way in.
2. Blow up the balloon with the air pump (but be careful not to overinflate it), and tie it off.
3. Grasp the balloon with an open hand at the stem end. The neck of the balloon should be in your palm and your fingers and thumb will grip the balloon down the sides.
4. While holding the balloon, palm down, turn it in a circular motion.
5. The goal is to get the penny to roll around inside the balloon (it may bounce around at first, but eventually it will calm down).
6. Once the coin begins spinning, use your other hand to stabilize the balloon. Your penny should keep spinning!

What's happening?

The real force in action here is called *centripetal force*, a center-seeking force. This means the force is always directed toward the center of the circle. The shape of the balloon is the initial element that makes the penny move in a circular path, and then *centripetal force* takes over and keeps the penny moving in a circular motion inside the balloon.

MORE: Try using other coins. Does it take longer to get them going? Do they spin for longer?

This experiment requires hammering a nail into a board, so be sure to ask for help from an adult!

Materials you'll need:

- Nails
- Hammer
- Flat wooden board

What to do:

1. Hammer a large nail far enough into a board so that the nail stands securely upright.
2. Lay a second large nail (we'll call this NAIL A) on the flat surface of the board and place another nail (NAIL B) across it with its head on one side of the nail (close to the body) and the rest of the nail on the other side.
3. Place another nail on NAIL A, in the same manner as NAIL B, but this time facing the opposite direction. The head should still be close to the body of NAIL A, but the point should be facing the opposite direction of NAIL B. However both this nail and NAIL B should have their heads close together, just on other sides of the nail.
4. Continue this pattern, alternating sides, until there is no more room for nails on NAIL A.
5. Now, place another nail on top of this entire assembly, head to tail with the NAIL A.

6. Carefully pick up the assembly and balance it on the
   upright nail.

What's happening?

In a gravitational field, any object is most stable when its center of mass (or center of weight) is low. The center of weight of this assembly when it's balanced. When it swings to the side, its center of mass rises, but gravity brings the assembly back into balance.

181)   RACQUET BALL EXPERIMENT

This experiment uses a sharp blade, so be sure to ask an adult for help.

Materials you'll need:

- 1 racquet ball
- Sharp knife

What to do:

1. With the help of an adult, use the sharp knife to cut a racquet ball into two halves.
2. Trim each half so that it is slightly smaller than a hemisphere.
3. Turn one side inside-out and drop it (bulge-side-up) on a hard surface.
4. What happens?

What's happening?

When you turn the hemisphere inside-out you create *potential energy* (stored energy) in the ball. When you drop it and it hits the hard surface, it turns potential energy into *kinetic energy* (moving energy) allowing the ball to bounce to a great height.

182) MAKE A BUBBLE INSIDE A BUBBLE

What's more fun than a bubble? A bubble INSIDE a bubble!

Materials you'll need:

- 1 Tablespoon granulated sugar
- 2 Tablespoons dish soap
- Water
- Tablespoon
- Scissors
- Pipette
- Cup

What to do:

1. Fill a cup with 9 ounces of water.
2. Add the tablespoon of granulated sugar to the water.
3. Now pour 2 tablespoons of dish soap into the mix and stir.
4. Use scissors to cut off the very end of the pipette bulb.
5. Dip your fingers in the solution and wipe the surface of table with it.
6. Dip the open bulb end of the pipette into the solution and use it to blow a bubble on the table.
7. Now dip the pipette back into the solution and place the bulb INSIDE the first bubble. Now blow a second bubble!
8. Keep going if you can, and see how many bubbles you can blow up inside.

What's happening?

The oily film you see is actually two separate layers of soap attached to hydrogen-bonded water. And the sugar you added makes those bubbles last longer.   When you're blowing a bubble inside of your first bubble, the first bubble expanded, because when you blow the second bubble, you are increasing the *volume of air* inside both of them! The bubbles can stretch because the hydrogen bonds of the water combined with the soap and sugar are very elastic.

183)   COLOR BLEACH

Bleach should always be handled with care, so be sure to ask for adult help with this experiment!

Materials you'll need:

- Water
- Bleach (ask for help!)
- Food coloring
- 2 clear cups

What to do:

1. Fill one of the cups ¾ full with tap water.
2. Add 2 drops of food coloring to the water and observe its vividness (brightness) in the water.
3. Now fill the other cup ¼ full with bleach and pour it into the cup of colorful water.
4. To mix, pour the mixture back and forth between cups 3-4 times.
5. Allow the new mixture sit for several minutes
6. What changed?

What's happening?

Bleach fades colors, through a process called *bleaching*, which happens through *oxidation or reduction*. Oxidation works by breaking the chemical bonds between the molecules of dye. The *oxidized molecules*, can no longer absorb visible light, which fades their colors. Reduction works by converting the dye's double bonds into single bond, which also prevents the molecule from absorb visible light.

184)   MAKE A SYMPHONY OF COLOR

Materials you'll need:

- A flat tray
- 3 different colors of food dye
- Whole milk (do not use low-fat!)
- Liquid dishwashing soap

What to do:

1. Carefully pour the milk into the tray so that it just covers the bottom
2. Add about 6-8 drops of different colored food coloring onto the milk in different spots
3. Add about 5 drops of the liquid soap onto the drops of food coloring and watch the show!
4. To clean up, simply pour the colored milk down the drain.

What's happening?

The main job of dish soap it to go after fat and break it down. There is fat in whole milk, so when you drop the soap into the tray, it tries to break down the fat in the milk. While it was doing that, it caused the colors to scatter and mix creating a very colorful display.

185)   COLOR CHEMISTRY

This experiment shows that magic is just science. In this demonstration you will see an almost clear liquid suddenly turn dark blue in a flash. This experiment uses iodine and should only be done with the help of an adult.

Materials you'll need:

- 3 clear plastic cups 4 ounces or larger
- A 1000 mg Vitamin C tablet
- Iodine (2%)
- Hydrogen peroxide (3%)
- Liquid laundry starch
- Safety goggles
- Measuring spoons
- Measuring cup

What to do:

1. Put on those safety goggles and crush the Vitamin C tablet into a fine powder by placing it into a plastic bag and mashing it with a rolling pin.
2. Put all the powder in the first cup and add 2 ounces of warm water. Stir for at least 30 seconds. Label it as LIQUID A.
3. Now put 1 teaspoon of LIQUID A into a new cup and add 2 ounces of warm water and 1 teaspoon of iodine. This will be LIQUID B.

4. In the last cup, mix 2 ounces of warm water, 1 Tablespoon of the hydrogen peroxide and ½ teaspoon of the liquid starch. This is LIQUID C.
5. Now pour all of LIQUID B into LIQUID C. Pour them back and forth between the two cups to mix them.
6. Place the cup down and watch.

What's happening?

This is an example of the chemical reaction known as the *Iodine Clock Reaction*. It is called a clock reaction because you can change the amount of time it takes for the liquids to turn blue. Starch is trying to turn the iodine blue, and Vitamin C stops it from turning blue. It's a chemistry battle!

186)   SHARPIE PEN EXPERIMENT

This colorful experiment uses rubbing alcohol so be sure to ask an adult for help.

Materials you'll need:

- Plastic cup with a large mouth
- Rubber band
- 91% isopropyl alcohol (get help with this!)
- Sharpie pens in different colors
- Piece of white cotton fabric (a t-shirt works great)
- Dropper squeeze bottle (such as an eyedropper or pipette)

What to do:

1. Place the plastic cup, upside down, in the middle of the white piece of cotton fabric.
2. Stretch the rubber band over the fabric and the cup to keep it in place.
3. Place dots of ink from one marker in a circle pattern about the size of a quarter in the center of the stretched out fabric. If you like, use another color marker to fill in spaces in between the first dots. There should be a quarter size circle of dots in the middle of the plastic cup opening when you are finished.
4. With adult help and the dropper, squeeze about 20 drops of rubbing alcohol into the center of the circle of dots. DO NOT POUR IT. The key is to DRIP the rubbing alcohol slowly in the center of the design.

5. Allow the design to dry for 3-5 minutes before moving on to a new area of the fabric.
6. If you are using this experiment to decorate a shirt, it is important to heat-set the colors by placing the shirt in the laundry dryer for about 15 minutes.

What's happening?
This is a lesson in *solubility, color mixing,* and the *movement of molecules.* The Sharpie markers contain permanent ink, which is *hydrophobic,* meaning it is *not soluble* in water. *Not soluble* means that it won't wash away in water. However, the permanent ink molecules are *soluble* in the rubbing alcohol. This carries the different colors of ink with it as it spreads in a circular pattern from the center of the fabric.

187)   WHAT COLOR IS BLACK?

This experiment will use chromatography to figure out if black is really black.

Materials you'll need:

- Round filter paper (or coffee filter)
- Scissors
- Small plastic cup
- Black maker (NOT a permanent maker, it needs to be water soluble)
- Pipe cleaner
- Small personal fan (with no front piece)
- Small plastic lid
- Strong glue
- Water
- Plastic pipette (or eye dropper)

What to do:

1. Use the strong glue to attach a plastic lid (make the filter piece the same size as the lid) to the fan blades with the edges of the lid pointing in an outward direction.
2. Turn the fan on low.
3. Lay the filter paper on the plastic lid, allowing the edges of the lid to hold it in place (this may take some time to get the paper spinning just right, so be patient and keep trying)

4. Touch the tip of the black pen to the spinning filter paper to draw a perfect circle in the very center of the filter paper.
5. Using the pipette, drip a few drops of water into the middle of the black circle. Centripetal force from the spinning paper should make the ink to spread out in a circular pattern.
6. What colors do you see?

What's happening?

The colors that you see on the filter paper proves that black a combination of colors. This technique of color separation is called *chromatography* (remember our **Candy Chromatography** experiment?) The ink dissolves in water (hence, *water soluble* marker) and moves in between the fibers of the paper where it is separated into bands of color. Depending on what kind of marker you use, you may see six or seven different circles of color.

MORE: Experiment with other *water-soluble* black pens in your home. How are they different?

188)   PRIMARY COLOR MIXING

How many different colors can you make using just red, yellow, and blue? Ask an adult for help with the warm water.

Materials you'll need:

- Plastic ice tray
- 3 plastic cups
- Red, yellow, and blue food coloring
- Warm Water (ask for help!)
- 3 Pipettes (or eye dropper)

What to do:

1. Fill the three plastic cups ¾ full with warm water.
2. Add yellow food coloring to one cup, blue to another, and red in the last one.
3. Dip the end of the pipette into the blue water, squeeze the bulb to push the air out, and release your squeeze to draw up some colorful liquid.
4. Squirt some of the blue liquid into one of the ice cube trays. Then add some yellow. What color did you make?
5. Now put some blue in another ice tray well. Now add red. What color did you make?
6. Keep mixing the colors in the ice tray until it's full of all different shades of colors!

What's happening?

Red, yellow, and blue are what's known as *primary colors*. The colors you make by mixing *primary colors* together creates *secondary colors*, such as orange, green, and yellow.

189)   INVISIBLE SODA

Materials you'll need:

- 20 oz. bottle of regular brown soda (like Pepsi, Coke, or Root Beer)
- Some 2% milk

What to do:

1. Open a fresh bottle of soda.
2. Pour out a little bit, and then pour some 2% milk into the bottle, until it is full to the very top.
3. Replace cap tightly.
4. Now you wait. Check on the bottle every 30 minutes or so, but be patient!

What's happening?

The milk settles at the bottom of the bottle, and the soda should become invisible. This happens because of a there is phosphoric acid in the soda and it reacts with the milk. *Phosphoric acid molecules* attach to the molecules of milk, separating them from the rest of the liquid, and sink down to the bottom because they are denser. The remaining liquids, which have none of the color, floats on top.

190) WAX ERUPTION

Create a bubbling wax eruption inside of a beaker and learn how magma works! This experiment uses some heat, so be sure to ask an adult for help.

Materials you'll need:

- Hot plate or Bunsen burner (ask for help!)
- Sand
- Wax
- Pyrex beaker (or other heat proof glass)
- Water

What to do:

- Place a piece of wax about 1" x 1" in the bottom the beaker, as close to the middle as possible.
- Pour enough sand into the beaker to completely cover the piece of wax.
- Slowly pour water into the beaker until the cup is 2/3 full.
- With the help of an adult, put the beaker onto the hot plate and turn the heat to medium-high.
- Observe what happens!

What's happening?

The very middle of the Earth (the core) is made up of very hot, liquid magma that occasionally rises to the surface to erupt out of the Earth's crust. Most of the eruptions (80

%!) occur underwater! With the Wax Eruption experiment, you "made" liquid hot magma by heating wax. The hot wax bubbles through the sand and makes mini-eruptions in the surface of the sand. Each bubble in the sand's surface represents and underwater volcano.

191) CHEMICAL CABBAGE

This experiment uses a blender, so be sure to ask an adult for help.

Materials you'll need:

- Some leaves of red cabbage
- A blender
- A strainer
- Plastic cups
- White paper
- Water
- Vinegar
- Laundry detergent

What to do:

1. First you must make your Indicator Cabbage Juice (warning, it smells awful!)
2. Peel off six red cabbage leaves and put them in a blender
3. Fill the blender halfway with water. Now liquefy!
4. Pour the cabbage juice through a strainer and into a large container to filter out the big chunks.
5. Now set out plastic glasses side-by-side, against a white piece of paper.
6. Fill each glass half full with cabbage juice.
7. Add a little vinegar to the first glass of cabbage juice.
8. Stir with a spoon and notice that the color changes to red, which indicates that vinegar is an acid.

9. In the second cup, add a teaspoon of laundry detergent. Note that the liquid turns green which indicates this chemical is a base. (Save these two glasses of red and green liquid for future reference).

10. Now add whatever substance you want to the last cup of cabbage juice and compare. Is your last substance an acid or a base? (Try orange juice, milk, salt, other soaps or lemonade).

What's happening?

As you have seen, the cabbage juice turns red when it is mixed with an acid (the vinegar), or green when it is mixed with a base (the detergent). An acid is the opposite of a base. Red cabbage juice is considered to be an *indicator* because it shows something about the chemical composition of other substances. If there is no color change at all, the substance that you are testing is considered *neutral* (water, for example).

192)   BALANCING ACT

This balancing act experiment seems to defy gravity, but it's just physics! This involves the use of knives, so be sure to ask an adult for help!

Materials you'll need:

- Drinking glass
- Three butter knives that are exactly the same in size and style
- Three empty glass bottles
- Water

What to do:

1. Place your three glass bottles an equal distance apart (they should be close enough that the knives can reach from the bottles' mouths to the center).
2. With an adult helper, intertwine the knives, where the blade of one knife lays atop the handle of another. This should make a triangle shape in between them.
3. Rest the knives on top of the bottles—the handles should sit on the mouths of the bottles. (The knives should stay intertwined).
4. Set a drinking glass on top of the knives at the spot where they all intersect.
5. Now pour water into the glass to see how sturdy your balancing act is!

What's happening?

The blades of the knives are intertwined in a *weave*. Each knife goes underneath a second knife while going over the top of a third. This makes it so when pressure is applied to all three knives (in this case, the glass full of water) *force* is being applied both downward and upward. The handles on the bottles, and the glass on the intersecting blade points creates *balanced forces* which allow the full glass of water to sit on your steady apparatus.

TOUGH TO BREAK TISSUE

In this experiment, you learn how to make tissue paper so strong you can't break it!

Materials:

- Tissue paper
- Toilet paper roll
- Wooden dowel (the handle of a wooden cooking utensil works great)
- Rubber band
- Salt

What to do:

1. Place the tissue paper over one end of the towel roll and secure it with a rubber band.
2. Now try to puncture the tissue paper by sticking the wooden dowel through the tube (you should be able to easily break the paper).
3. Repeat step 1 to set up the experiment again.
4. Set the toilet paper roll on a table so that the tissue paper side is lying flat against the surface.
5. Pour salt into the roll until it is ¾ full.
6. Lightly tap the roll on the flat surface to pack the salt.
7. Now try pushing the dowel through the tube and into the tissue paper.
8. What happens?

What's happening?

As you may have guessed, the key to the tissue paper's strength is the addition of salt. All the grains of salt (hundreds of thousands) increase the *surface area* which disperses the force of the dowel. The force spreads from grain to grain, and also spreads evenly across the *surface area* of the tissue paper. The increased *surface area* makes the tissue paper seems stronger than it was before.

193)   PING PONG BLASTER

This ping pong popper toy will teach you about *combustion,* but be sure you ask an adult for help before you start this experiment.

Materials you'll need:

- Plastic car battery filler
- Spark igniter
- 91% rubbing alcohol
- Philips screwdriver

What to do:

1. Undo the top of the battery filler; this is the part you will be using, so don't worry about the tube.
2. Use your screwdriver to make a hole near the center of the filler.
3. With adult help, carefully take apart the spark igniter handle, removing the nut and washer too.
4. Insert the igniter through the opening in the plastic filler, and push it through until you can see the threading.
5. Now, reassemble the igniter (it helps to hold onto the igniter from the inside), and make sure the nut and washer are tightly secured.
6. Test the igniter to make sure it works
7. Pour 10-15 drops of rubbing alcohol inside of the filler.
8. Insert the ping pong ball into the filler opening, and make sure it is snugly in.

9. Point your blaster away from you and ignite for blast
    off!

What's happening?

What makes the ping pong ball fly? A little thing called
combustion! This occurred when the spark igniter lit the
vapors of the rubbing alcohol inside the popper, which
caused the air inside to expand very quickly. At that point
the rapidly expanding air is looking for a way to escape the
blaster (the *point of least resistance*). In this experiment,
the point of least resistance is the seal between the ping
pong ball and the popper

XVII. ADVANCED SCIENTIFIC EXPERIMENTS

194)   CREATE A FOG TWISTER!

This cool fog tornado experiment involves a lot of tricky parts and dry ice, so be sure to ask for adult help with this one!

Materials you'll need:

- A large cardboard box (paper copier boxes work great!)
- Exact-o knife (ask for help!)
- A small, 12-volt computer fan. This should be at least 3×3 inches. (you can find these at an electronics or computer store)
- A piece of clear plastic 10x17 inches (clear plastic wrap also works).
- A small plastic food container
- Dry ice (ask for help!)
- Tape
- Black paint
- Optional: battery-powered tap light

What to do:

1. Ask an adult for help with this part. Following the diagram below, have your adult helper cut away the marks in gray.

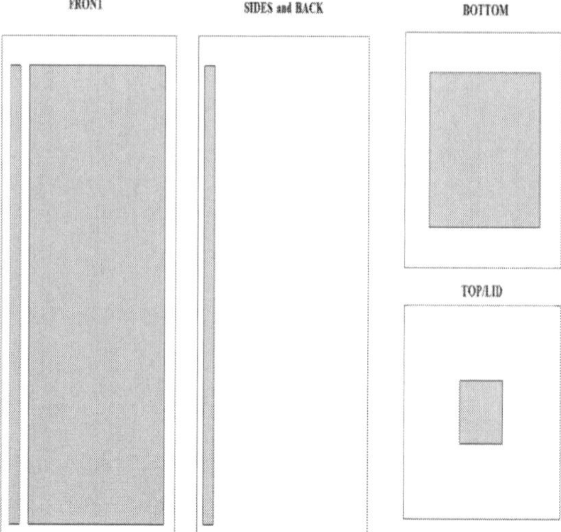

2. The opening at the top should be just a tad smaller than the size of your fan.
3. The opening at the bottom should be bigger than the plastic food container.
4. Paint the inside of the box with black paint and allow it to dry.
5. Tape your clear pieces of plastic to the large window on the front side of the box. (But be sure to keep the thin slot above it open—air will need to flow through it).
6. If you have the optional tap light, attach it now to the inside of the lid, close to the opening for the fan. Turn it on.
7. Now it's time to make your Fog Tornado!
8. Plug in your mini fan and place it over the small hole on the lid, with the breeze blowing up. Place the lid on the box.

9. Have your adult helper place some small chunks of dry
    ice in the plastic food container (be sure they are
    wearing gloves!)
10. Now add a little warm water to the food container (this
    will start creating fog).
11. Take the whole box and place it over the container of
    "fog" and watch your cool tornado go!

What's happening?

As the fan pulls air out of the box, air from outside the box
is sucked in through the slots on the sides. The way the
slots are positioned causes the incoming air to create a
vortex (spiral) of air as it is sucked in, and then drawn up
and out of the box.

MORE: Play around with your tornado box and see what
affects the vortex of air! Try to control the shape of the
tornado by covering up some of the slots, or try a bigger
box to see if you can make a bigger tornado!

195)   BUILD AN ELECTRIC MOTOR

This advanced experiment will show you how a motor changes electrical energy into motion.

Materials you'll need:

- 3 feet of 22-gauge or 24-gauge solid-core insulated wire
- 2 disk magnets
- 2 insulated test cables with a clip on each end
- a plastic cup
- two large rubber bands
- two jumbo size (2-inch) paper clips
- D-cell battery
- wire strippers
- permanent marker

What to do:

1. Take the 3-foot piece of insulated wire. Starting about 3 inches from the end of the wire, wrap it seven times around the D-cell battery to form a coil. Wrap the ends of the wire a couple of times around the coil to hold it together.
2. Use the wire strippers to remove the insulation from the two ends of the coil.
3. Take the two large loops of the paper clips and straighten them.
4. Turn the plastic cup upside down and place a magnet on top in the center.

5. Attach the other magnet inside the cup, directly beneath the original magnet (this will create a stronger magnetic field).
6. Wrap the two large rubber bands around the base of the cup.
7. Insert the straightened paper clips into the rubber bands, so they stand upright over the bottom of the cup.
8. Rest the ends of the coil in the cradles formed by the paper clips. Adjust the height of the paper clips so that when the coil spins, it is just above the magnets, but doesn't touch them. (Adjust the coil and the clips so that the coil is well balance and spins freely. This is important for the experiment to work).
9. Trim off any excess wire that sticks through the paper clips.
10. Attach one of the clip cables to each paper clip, just above the rubber bands. (You may need to adjust the clips to make sure the coil still spins freely).
11. Hold the other ends of the clip leads against the two poles of the D-cell battery. (The coil should rotate to an almost horizontal position).
12. Now, remove the coil from the paper clips. Hold it vertically and use the marker to color the top half of one of the two end wires. Allow the ink to dry then apply a second coat. Wait for it to dry again, then hang the coil back on the paper clips.
13. Connect it back up to the D-cell battery, and give the coil a gentle spin. It should keep spinning on its own!

What's happening?

The electric current from your battery and coil produces a magnetic field. This magnetic field in the coil aligns itself with the magnets. This is why it turns horizontal when you first turn it on. When you "paint" one of the bare wire ends of the coil, it "turns off" the current, which allows the coil to halt aligning itself with the magnet, and then re-align itself, over and over again, which makes it spin!

196)   MAKE A CONDUCTIVITY TESTER

Pure water does not conduct electricity very well. However, when certain substances are dissolved in water, the solution does conduct electricity. You can make a simple device that shows how well a solution conducts electricity. This device uses a flashlight bulb to indicate how well the solution conducts electricity. The better the solution conducts electricity, the brighter the bulb will glow.

Materials you'll need:

- a 12-volt AC adapter
- an audio cable with a 1/4-inch or 1/8-inch monaural plug on one end
- a 12-volt flashlight bulb and socket
- a block of wood about 4in x 4in x 1in (The electrical connections will be made on this block)
- two 1-inch wood screws (these hold the lamp socket to the block of wood)
- one 3/4-inch round-headed screw and washer (these will be used to make an electrical connection)
- wire cutter and wire stripper
- a screw driver

What to do:

1. Cut the plug from the end of the cord of the AC adapter. Separate about four inches of the cord into its two conductors. Remove about 1 inch of insulation from each of the conductors.

2. Cut the cord of the audio cable about 2 feet from the plug. Remove about four inches of insulation from the cut end of the cable. This will expose bare stranded wire wrapped around insulation that covers a center wire. Unwrap the stranded wires from the insulation and twist the strands together to make a single bundle. Strip about 1 inch of the inner insulation from the center wire.

3. Use wood screws to attach the lamp base (the socket) to the block of wood.

4. Put the washer on the round-head screw and screw it into the block next to the lamp base (but don't tighten the screw yet)

5. Wrap one wire from the AC adapter around the screw above the washer. Wrap the end of the bundled wire from the audio plug around the same screw. Tighten the screw to fasten the two wires together.

6. Attach the remaining wire from the AC adapter to one of the terminals of the lamp base. Attach the remaining wire from the audio plug to the other terminal of the lamp base.

7. Screw the 12-volt flashlight lamp into the lamp base.

8. It is now ready to use—to see if it works, plug the AC adapter into an AC outlet. The lamp should not light. Touch the audio plug sideways to a coin. When the two metal conductors of the plug are shorted by the coin, the lamp will glow brightly. (The glow indicates that current is easily flowing through the coin).

9. Now, to use it! Put some tap water into a cup. Insert the end of the audio plug into the water, the light should glow a little bit, which means the tap water conducts electricity but poorly.

10. Now add some table salt to the water and stir the
     mixture. The lamp will glow brightly when the plug is
     put into the solution, because salt solution is a great
     conductor of electricity.
11. Try other materials around your house to see how well
     they conduct electricity when mixed with water—sugar,
     shampoo, laundry detergent, fizzy tablets, rubbing
     alcohol, and baking soda are just a few, but basically
     anything that dissolves in water ca be tested. Just be
     sure to rinse the plug in water and dry it before testing
     different solutions. Also be sure to only keep the plug
     in the solutions for 10-15 seconds.
12. Record what conducts electricity well, what conducts
     poorly, and what doesn't conduct at all.

What's happening?

An electric current is a flow of electrical charge. When a
metal conducts electricity, the charge is carried by
electrons (negatively charged particles) moving through the
metal. When a solution conducts electricity, the charge is
carried by *ions* (atoms that have an electrical charge)
moving through the solution. Some ions have a negative
charge and some have a positive charge.

Pure tap water contains very few ions, so it does not
conduct electricity very well. When you dissolve table salt
in tap water, you add ions, which helps the solution to
conduct electricity. The ions come from the salt (aka
sodium chloride). Sodium ions have a positive charge, and
chloride ions have a negative charge. Because sodium
chloride is made up of ions, it is called an *ionic substance*.

In your experiments, you notice that not all substances are made up of ions. Some are made of *uncharged particles* called *molecules* (sugar is an example).

197)   MAKE A SMOKE BOMB

This advanced experiment uses definitely requires the help of an adult.

Materials you'll need:

- Regular sugar
- Potassium nitrate, also known as saltpeter (you can find this in some garden stores (fertilizer section) or online)
- Pan
- Stovetop
- spoon
- aluminum foil
- well-ventilated area outside
- lighter

What to do:

1. Pour about 3 parts potassium nitrate to 2 parts sugar into the pan (just make sure you have more potassium nitrate than sugar)
2. Apply low heat to the pan and stir the mixture using long strokes of a spoon. (If sugar starts to melt along the edges of the pan, reduce the heat!) The mixture should melt and become a caramel or chocolate color.
3. Remove the pan from heat once everything is liquefied.
4. Pour the liquid onto a piece of foil. (You can pour the smoke bomb into a cookie pattern shape or mold. The shape and size WILL affect how it burns, though).

5. Allow the smoke bomb to cool, and then pull it off the foil
6. Take it outside and light it!

What's happening?

When Potassium Nitrate (KNO_3) is heated it releases oxygen and becomes KNO_2. The released oxygen combined with the flame results in the rapid combustion of the sugar, which makes your awesome smoke bomb.

198)   BOWLING BALL EXPERIMENT

Did you know not all bowling balls sink in water? Do this experiment to find out why!

Materials you'll need:

- 11 pound (or lighter) bowling ball
- 13 pound (or heavier) bowling ball
- A bathtub (or large trashcan, or large aquarium)
- bathroom scale
- water

What to do:

1. Fill the bathtub ¾ full with water.
2. Carefully place (do not drop) one bowling ball in the water. Does it float or sink? Write down the weight of the bowling ball and how it behaved.
3. Repeat this experiment, noting the weight of each bowling ball and how it behaves.
4. What did you discover?

What's happening?

It seems that anything heavier than 12 pounds will sink, but bowling balls less than 12 pounds float. But why? Regulation bowling balls range from about 8 to 16 pounds, but they are all the same size around. The rules of bowling state that the circumference of a ball (the distance around it) should be 27 inches, so we'll use that number to calculate density.

In science, you use metric measurements, so we need to convert inches to centimeters using the conversion factor of 1 inch = 2.54 centimeters (cm). The circumference of the bowling ball in centimeters is 27 inches x 2.54 cm/inch = 68.58 cm. Now you can determine the volumes of your bowling balls using the equations for circumference and volume of a sphere. The equation looks like this: circumference = 2 π R. Once you have solved for the radius of each ball, you can find the volumes using the equation: volume = 4/3 π R^3. Bowling balls do have 3-4 holes in each ball for the fingers. However, it's fair to say the holes do not affect the overall density of the bowling balls enough to change the experiment. So, we'll use the number 5,452 cm^3.

Now to find the weight: weigh yourself, then weigh yourself holding a bowling ball, and subtract the two weights to find the weight of the ball. And again, you'll have to convert the weight from pounds to grams. The conversion between pounds and grams is 1 lb. = 453.6 g. Finally, to get the density of your bowling balls, divide the weight of you ball (in grams) by the volume (in centimeters). The equation looks like this: Density= mass/volume. If the final amount is less than 1.0, it is less dense than water. If it is more, it is *denser* than water.

199)   BUILD A REAL HOVERCRAFT YOU CAN RIDE!

The initial instructions for this awesome experiment came from the Jimmy Kimmel Live Show. Enjoy, be safe, and HAVE FUN! This experiment definitely requires the help of an adult, so let them in on the fun, too!

Materials you'll need:

- 4' x 4' ¾ -inch piece of plywood
- 5' x 5' piece of heavy-duty tarp material
- .25 inch machine bolt (1.5 inches long) with nut
- 2 2-inch washers for the bolt
- Plastic cover for a round electrical box
- Leaf blower (cordless electric leaf blowers are the best)
- Optional: Lawn chair
- String
- Pencil
- Duct tape
- Jig saw (adult use only!)
- Staple gun (adult use only!)
- Drill (adult use only!)
- Utility knife (adult use only!)
- Sandpaper
- Optional: Hot glue gun and foam pipe insulation
- A helmet

What to do:

1. Mark the center of the plywood by drawing lines between opposite corners.
2. Nail a small nail into the center of the panel and use a pencil tied to a piece of string as a compass to draw a 4' circle around the nail.
3. Using a jig saw (with an adult!) carefully cut the circle out of the panel and sand the edges smooth with sandpaper.
4. Measure the width of the leaf blower output and use the jig saw to cut a hole the same width (it should be about 2 feet from the center of the plywood and it should fit snugly. If needed, use duct tape around the blower to tighten the seal).
5. If you want to attach the optional seat to the top, do it now. Drill holes and attach with machine screws. Install the machine screws through the bottom, and counter-sink all screws.
6. Lay the plywood on the center of the tarp and wrap the tarp over the edge. Staple the tarp onto the top of the board with a staple gun along the edge so that the staples almost touch each other. Keep the tarp taught while stapling. When finished, cut off any extra tarp material
7. Drill a ¼ inch hole through the center of the plywood and tarp.
8. Drill a ¼ inch hole through the center of the electrical box cover.
9. Feed a bolt through a washer and the plastic electrical box cover, and then up through the bottom of hovercraft. Secure with a washer and the nut. (Cover the screw head with duct tape)

10. Use a utility knife to carefully cut 6 evenly spaced holes into the tarp material in a circle about 10 inches from the center. The holes should be about 1.5 inches wide. (If you need help, use a quarter to trace the circles.)
11. Optional: To make bumpers, you can cut foam pipe insulation and attach to the outside edge with hot glue.
12. You're now ready to use it! Place the hovercraft on a smooth surface, such as concrete or asphalt. Center yourself on top of the hovercraft, or sit in the chair. Put on your helmet.
13. Power up the leaf blower and have someone give you a push (who can also help guide your hovercraft, since there are no brakes or steering).
14. You're hovering!

What's happening?

The leaf blower provides *propulsion*, which makes you move. This *moving air* is what allows planes to get off the ground. When you turn on the leaf blower, it forces air out the holes of the tarp and towards the edge of the hovercraft. While you might think it would be difficult for a small blower to lift over you, it is not. The leaf blower produces a large volume of moving air, which is spread out over a relatively thin area, so it has a great amount of lift. This layer of air reduces *friction* and allows you to hover along on a cushion of air.

Made in the USA
Middletown, DE
14 December 2015